Intelligent Control and Learning Systems

Volume 15

Series Editor

The Springer book series Intelligent Control and Learning Systems addresses the emerging advances in intelligent control and learning systems from both mathematical theory and engineering application perspectives. It is a series of monographs and contributed volumes focusing on the in-depth exploration of learning theory in control such as iterative learning, machine learning, deep learning, and others sharing the learning concept, and their corresponding intelligent system frameworks in engineering applications. This series is featured by the comprehensive understanding and practical application of learning mechanisms. This book series involves applications in industrial engineering, control engineering, and material engineering, etc.

The Intelligent Control and Learning System book series promotes the exchange of emerging theory and technology of intelligent control and learning systems between academia and industry. It aims to provide a timely reflection of the advances in intelligent control and learning systems. This book series is distinguished by the combination of the system theory and emerging topics such as machine learning, artificial intelligence, and big data. As a collection, this book series provides valuable resources to a wide audience in academia, the engineering research community, industry and anyone else looking to expand their knowledge in intelligent control and learning systems.

Zhengle Zhang · Tiedong Ma · Bing Cui

Distributed Impulsive Coordination of Multi-Agent Systems

Springer

Zhengle Zhang
Henan University of Technology
Zhengzhou, China

Tiedong Ma
Chongqing University
Chongqing, China

Bing Cui
Beijing Institute of Technology
Beijing, China

ISSN 2662-5458 ISSN 2662-5466 (electronic)
Intelligent Control and Learning Systems
ISBN 978-981-97-6261-3 ISBN 978-981-97-6262-0 (eBook)
https://doi.org/10.1007/978-981-97-6262-0

This work was supported by National Natural Science Foundation of China (U22A20101, 62173051, 62103066), Chongqing Technology Innovation and Application Development Special Key Project (CSTB2022TIAD-CUX0015, CSTB2022TIAD-KPX0162) and Startup Foundation for Introducing Talent of Henan University of Technology (0004-31401590).

This Springer imprint is published by the registered company Springer Nature Singapore Pte Ltd.
The registered company address is: 152 Beach Road, #21-01/04 Gateway East, Singapore 189721, Singapore

To My Family

Zhengle Zhang

To My Family

Tiedong Ma

To My Family

Bing Cui

Preface

In recent years, the coordination control for multi-agent systems (MASs) has been paid more and more attention by researchers, which contains consensus problem and formation problem. The continuous control method needs to obtain the state information of the system continuously in the whole control process, which results in large information transmission consumption. Compared with continuous control, impulsive control as a discontinuous control method, only needs to exchange information at impulsive instants, which can greatly reduce the transmission consumption in the communication process. Based on the above considerations, in order to effectively reduce the communication cost, this book adopts the impulsive control method to achieve the consensus of nonlinear MASs. In practice, MASs are often affected by environmental and hardware conditions, resulting in the limited application of impulsive control, such as the impulsive control signal is difficult to precisely apply at a fixed time, the exist of impulsive control gain error, and it is difficult to control all nodes at the same impulsive instant. Moreover, the actual MASs model often has unknown parameters, stochastic disturbance, and time delay, and the limitations in these models will increase the difficulty of controller design. In addition, stochastic switching topologies and cyber attacks will limit the communication between nodes in MASs, which will greatly affect the coordination control performance. On the basis of the existing research results, this book considers the above limitations in practice and studies the key problems of distributed impulsive coordination control for MASs. The main work of this book can be summarized as follows:

1. The variable impulsive consensus problem of MASs is studied in view of the difficulty of precise application of impulsive control signals at fixed time. Firstly, a distributed impulsive controller is designed based on graph theory knowledge. Secondly, two control strategies, impulsive time window and odd impulsive sequence, are proposed. Then, a novel variable impulsive control strategy is proposed, in which the impulsive signal only needs to be applied in a certain range to achieve the control objective, which can greatly improve the practicability and effectiveness of impulsive control. Then, based on the impulsive differential equation and Lyapunov stability theory, the sufficient conditions for

achieving consensus are derived. Finally, the variable impulsive control method is extended to the intermittent impulsive control method, which allows the impulsive controller to not work for a period of time, so as to further improve the flexibility of the control strategy.

2. Aiming at the situation that all agents in MASs cannot be controlled at one impulsive instant, the asynchronous impulsive consensus problem of MASs is studied. Firstly, the asynchronous impulsive controller is designed, that is, the consensus can be achieved by controlling some agents at one impulsive instant. Then, based on the impulsive differential equation and Lyapunov stability theory, the sufficient conditions for realizing the asynchronous impulsive consensus are derived. Finally, two schemes are proposed to reduce the conservatism of the asynchronous impulsive control method and enhance the practicability of the control protocol.
3. The consensus of a class of nonlinear delayed MASs with impulsive control is addressed. By combining the Lyapunov stability theory, graph theory, impulsive differential system, stochastic system, and mathematical induction approaches, some novel criteria on stochastic synchronization with impulsive effects are established, which is more rigorous and effective in practical systems.
4. The adaptive impulsive consensus problem of uncertain MASs with stochastic disturbance is studied. Firstly, the distributed impulsive controller and the discontinuous adaptive law are designed. The control strategy only needs to transmit state information at impulsive instant, which can reduce the communication cost. Then, based on the Lyapunov stability theory, impulsive differential equation theory, and infinite series correlation properties, the conditions of impulsive control gain and impulsive interval under the discontinuous adaptive impulsive control law are derived to ensure the consensus of MASs.
5. The consensus of fractional-order nonlinear MASs with control gain error via impulsive control protocol is investigated. Based on the Lyapunov stability theory for fractional-order system and impulsive system, the sufficient condition is given to realize the impulsive consensus of fractional-order nonlinear MASs. In addition, the control gain error is considered, which corresponds more to reality.
6. The impulsive consensus of MASs under cyber attacks is studied. (1) For the case of Denial-of-Service (DoS) attacks on MASs, the impulsive consensus of MASs is studied. Firstly, considering that DoS attacks will interrupt the communication of some edges in MASs, a distributed impulsive controller under DoS attacks is designed. Then, based on the recovery mechanism of DoS attacks, the control process is divided into three different windows, namely normal window, attack window, and regulate window. Then, based on the theories of impulsive differential equations and the Lyapunov stability, sufficient conditions are obtained to realize the consensus of MASs under DoS attacks. Finally, based on the obtained sufficient conditions, the selection methods of impulsive interval in three different windows are given. (2) For the case of deception attacks in MASs, the consensus of MASs is studied. Firstly, the deception attacks are considered in the controller-to-actuator channel and modeled as a Bernoulli distribution. In

addition, two kinds of deception attacks are considered, namely injection attacks and replacement attacks, then based on the theories of the Lyapunov stability, impulsive differential equation, and stochastic system, some sufficient conditions are derived to achieve the consensus of MASs under deception attacks where the mean-squared consensus errors are bounded, and the expression of the error bound is obtained simultaneously. Finally, based on the obtained sufficient conditions and the expression of the error bound, criteria for selecting control parameters, such as lower and upper bound of impulsive interval, impulsive control gain, and convergence rate are discussed to obtain better control performance.

7. The formation control problem of MASs with switching topology and impulsive input delay is studied. The impulsive control algorithm is designed, and the inequality condition of sampling time satisfying the formation is given by combining the impulsive differential equation. Based on the Hurwitz stability theory and relevant lemma, the control protocol and error system are analyzed.

This publication is a research reference whose intended audience includes researchers and postgraduate and graduate students.

Zhengzhou, China
Chongqing, China
Beijing, China
May 2024

Zhengle Zhang
Tiedong Ma
Bing Cui

Acknowledgments

The material of this book is based on our works in recent years. Over the years, the authors have benefited from discussions with our colleagues at the School of Automation, Chongqing University, and the College of Electrical Engineering, Henan University of Technology. Even though it is not possible to name all of them, the authors would like to particularly thank Xiaojie Su, Jiangshuai Huang, Fangzheng Xue, Shaoxin Sun, Caizhang Wu, Li Wang, Shijie Zhang, and Jingjing Yan, for their valuable suggestions, constructive comments, and support.

Thanks to the graduate students at the School of Automation, Chongqing University, Teng Li, Tiantian Yu, Wei Luo, Kun Li, and Yongrui Xiao. The authors are especially grateful to their families for their encouragement and never-ending support when it was most required. In addition, we would like to thank the editors at Springer for their professional and efficient handling of this project.

This work was supported in part by the National Natural Science Foundation of China under Grant U22A20101, Grant 62173051, and Grant 62103066; in part by the Chongqing Technology Innovation and Application Development Special Key Project under Grant CSTB2022TIAD-CUX0015 and Grant CSTB2022TIAD-KPX0162; in part by the National Natural Science Foundation of Chongqing under Grant CSTB2023NSCQ-MSX0452; and in part by the Startup Foundation for Introducing Talent of Henan University of Technology 0004-31401590.

Contents

Chapter 1
Introduction

1.1 Problem Background

Over the past few decades, multi-agent systems (MASs) have come to an attention of researchers inspired by the cooperative behavior observed in nature, such as the migration of birds in the sky, large groups of small fish for hunting and defense, and large groups of penguins for protection against the cold. With the development of society, many more demanding tasks cannot be accomplished by a single agent. The MASs are composed of multiple single agents, which can communicate, interact, and cooperate with each other to solve complex problems in a more efficient way and complete complex tasks that cannot be completed by a single agent.

In recent years, with the trend of social informatization becoming more and more obvious, MASs have been more and more widely used in life, industry, and science and technology. With the development of scientific research, the current research on MASs mainly includes consensus (Hu et al. 2023; Jiang et al. 2024; Lu and Jia 2023), formation (Zhou and Chen 2024; Yao et al. 2022; Cheng and Huang 2022), clustering (Wang et al. 2023; Xiao et al. 2023; Yu and Yu 2020), containment problem (Lu et al. 2024; Wang et al. 2021; Liu et al. 2021), and so on. Consensus control problem of MASs means that the state of all nodes in the system tends to be consistent by applying appropriate control inputs. As the basis of studying MASs, the consensus problem has been paid more and more attention by researchers.

The consensus of MASs is widely used in engineering fields such as robots, unmanned aerial vehicles (UAVs), and smart power grids. For example, in the multi-robot system, each robot is regarded as an agent. When multiple robots work together, it is necessary to use the consensus control algorithm to ensure efficient communication between robots and achieve collaborative work. Moreover, in the problem of uniform speed flight movement of multi-UAV formation, each UAV is regarded as an agent, and it is particularly important to keep the speed consensus among the agents, and the position also needs to meet a certain formation. Therefore, the consensus theory is the basis of the study of UAV cooperative control. In a smart grid system,

Z. Zhang et al., *Distributed Impulsive Coordination of Multi-Agent Systems*, Intelligent Control and Learning Systems 15, https://doi.org/10.1007/978-981-97-6262-0_1

various types of components such as generators, loads, and energy storage units are connected to each other, similar to the agents in MASs. The consensus control algorithm of MASs provides a theoretical basis for the research of current coordination control in smart grid. Therefore, it is of great engineering significance to study the consensus of MASs.

At present, the research on MASs mainly includes two control methods: centralized control and distributed control. The centralized method requires a central controller, but when the central controller fails, it has the potential to cause the break of the whole system, and it is difficult to meet the stable and efficient operation requirements of complex large-scale systems. Therefore, it is very expensive and even unrealistic to use centralized control method to control large-scale MASs. In order to make up for the deficiency of centralized control, it is necessary to adopt the method of distributed control, which only uses the neighbor information of the nodes in the MASs to change its own state according to the control target under the action of the controller, so as to complete complex or difficult tasks. When the distributed control method is adopted, the failure of one agent in the system will not lead to the break of the whole system (Sinafar et al. 2020; Cong et al. 2023; Li et al. 2023a). Therefore, the distributed control method is more robust and is considered by scholars to be more valuable for research. In engineering, taking the current coordination control of the smart grid as an example, when designing the distributed consensus control algorithm, each power generation unit in the grid can be regarded as an agent, which only needs to transmit real-time current information to the neighbor agent and does not need a central controller to receive the current information of all agents. With the distributed control method, even if a power generation unit in the DC grid fails, the whole grid system will not stop operation. Therefore, the distributed control method is of great significance in practical engineering.

In practice, the models of most MASs have nonlinearity, such as robots, drones, and components in smart grids. The models of these systems often need to be constructed through nonlinear equations (Huang et al. 2020; Wang et al. 2024). Compared with linear systems, the most typical characteristic of nonlinear systems is that they do not satisfy the superposition principle, which increases the difficulty and complexity of analysis. This book mainly studies a class of nonlinear MASs that satisfy the Lipschitz condition.

There are many phenomena in the real world that involve instantaneous changes, such as bank interest rates and biological populations under sudden environmental changes. This kind of change is very rapid, even in an instant, and the phenomenon of this sudden change can be described by the impulsive differential equation (Samoilenko 1995). The impulsive differential equation can describe the state of the system in detail and accurately in mathematics. Impulsive control is a kind of discontinuous control method based on impulsive differential equation, which can make the state of the system change at the appropriate instant (impulsive instant), so as to achieve the control objective. In some practical scenarios, impulsive control is more significant than continuous control, for example, continuous control cannot realize the regulation of bank savings interest rate. Moreover, in a chemical reaction, by adding reagents to the reactants, the state of the reactants is mutated, etc. In addition, since impulsive control only requires the system to transmit information at discrete times, it can save control losses to a large extent and also improve the confidentiality and security of the communication process. In recent years, impulsive control has attracted wide attention due to its simple structure, high confidentiality, strong robustness, and low control cost, and many researchers have carried out research in different fields, such as aerospace, image encryption, and ecological balance.

The impulsive consensus problem of MASs is also applied in practical engineering. Taking the current coordination control of DC microgrid as an example, each power generation unit in the grid is regarded as an agent. If the continuous control strategy is adopted, each power generation unit needs to maintain real-time communication with its neighbors, which requires a large amount of information transmission consumption and increases the work burden of communication equipment in the grid. Therefore, how to reduce the communication time between agents on the current coordination control of DC microgrid is a very valuable problem. By adopting the distributed impulsive consensus control method, the current distribution regulation in microgrid can be realized only by using discrete communication (Liu et al. 2018a).

It is worth noting that most of the existing research results on MASs consensus control usually consider the ideal situation, but in practice, MASs are often subject to various limitations. In practice, due to the environment and hardware factors, the impulsive control cannot be precisely applied at a fixed time, or all nodes cannot be controlled at the same impulsive instant, and it may be affected by the control gain error. Moreover, there are often limitations related to the system model in practice, such as unknown system parameters, random switching topology, time delay, and stochastic perturbation. In addition, the process of information transmission between nodes in MASs is often limited, and cyber attacks and communication delay are important factors leading to communication limitations. Therefore, when designing a distributed control strategy, the above limiting factors should be considered to enhance the practicability and feasibility of the control strategy.

1.2 Overview of the Multi-agent Systems

1.2.1 Consensus Problem

Recently, the consensus of MASs has become a hot research issue. Consensus problem refers to the fact that the state of all the following nodes can eventually be consistent with that of the leader node. Olfati-Saber and Murray (2004) first constructed a theoretical framework for the consensus problem of MASs, and Ren and Beard (2005) indicated that topological graph theory is the key to study the consensus problem of MASs.

The consensus problem of MASs is divided into two problems: regulation problem and tracking problem. The former refers to the design of a distributed controller for all nodes in the system, so that the states of all nodes tend to a constant or time-varying value. For the tracking consensus problem with a leader, the leader does not receive signals from other nodes, and the trajectory of the leader can be regarded as the expected trajectory, under the effectiveness of the distributed controller, the trajectory of all the following agents is consistent with the expected trajectory, that is, the consensus of all the following nodes and the leader is achieved. It is important to note that the leader does not send signals to all followers, but only to its neighboring follower nodes. In addition, the leader in the tracking conformance problem can be either a practical system or a virtual reference system designed according to control objectives.

So far, there have been many valuable control methods for achieving the consensus of MASs, such as adaptive control, sliding mode control, finite/fixed time control, and so on. For the adaptive control, Yao et al. (2020) realized the leader-following consensus of uncertain MASs, Li and Wang (2023) realized the adaptive output consensus of heterogeneous MASs via distributed dynamic compensator, and Wang et. al achieved the consensus of Lipschitz nonlinear uncertain MASs, Wu et al. (2024) investigated the event-triggered adaptive control problem for a class of leaderless MASs with output constraints and unknown dead zones. For the sliding mode control, Yu et al. (2021) designed an adaptive sliding mode fault-tolerant controller for first-order and second-order MASs to ensure the consensus can be realized. Li et al. (2023b) studied sliding mode control based on reinforcement learning to achieve the consensus of nonlinear uncertain MASs. Guo et al. (2023) designed a fully distributed sliding mode controller with fault compensation to deal with the consensus problem of discrete-time first-order MASs with actuator faults. For the finite/fixed time control, Dong et al. (2021) studied the finite-time tracking problem of on-strict feedback MASs. Du et al. (2020) realized the fixed time consensus of heterogeneous MAS. Cong et al. (2023) considered the constrained finite-time consensus control problem for second-order MASs with leaderless and leader–follower.

As mentioned above, there have been a lot of results on the issue of consensus problem. However, most of these achievements use continuous control methods, which require information transmission and communication at every moment of the control process and need to exert control input uninterruptedly over the system. Compared with continuous control, impulsive control only needs to transmit information in discrete control instants, which can save communication resources.

1.2.2 Formation Problem

Formation control of MASs is to design a controller so that all agents can obtain the required geometric formation, which is widely used in multi-robot cooperative combat, UAVs, and industrial production. Due to the extensive application background of formations, scholars have proposed many control methods, such as leading methods (Madhevan and Sreekumar 2013), behavior-based methods (Lee and Chwa 2017), and virtual structure methods (Pan et al. 2019) to solve the formation control problem of MASs. The proposed nonlinear consensus protocol of the time-invariant, time-varying, and trajectory tracking formation control is applied in Xiao et al. (2009). In Liu et al. (2020), the formation of a group of quad-rotors was asymptotically obtained eventually by using variants of a control protocol. Chen et al. (2008) investigated the formation problem of MASs which are represented by the first and second order differential equations. Li et al. (2014) adopted robust output regulation to study the formation problem of nonlinear MASs. Do (2014) completed the aim of formation by using the Lyapunov direct method and the backstepping control. Zhang et al. (2019) introduced a modified adaptive sliding mode in the undirected multi-spacecraft to address the formation problem. Nair et al. (2018) introduced sliding mode control for nonholonomic robots. Do (2008) investigated the formation of unicycle-type mobile robots via output-feedback control. Nascimento et al. (2013) studied active target tracking with the model predictive control method. Yu et al. (2018) studied the MASs by robust guaranteed cost time-varying formation tracking control.

It should be noticed that the approaches mentioned above are all continuous control methods to address the formation problem, as well as the state information needs to be derived from systems of all the response time, which results in more consumption of information transmitted between the agents of MASs. By contrast, the impulsive control method can greatly reduce the state information transmitted from all the systems as it only responses at impulsive instants.

1.3 Overview of Impulsive Control

Since the 1950s, researchers have noticed the existence of impulsive phenomenon in practice. Since the state of the system changes rapidly in an instant, the phenomenon cannot be described by continuous differential equations, then researchers further proposed the concept of impulsive differential equations. In the control domain, impulses can be used as both disturbance and control input. When the impulse is a disturbance, it is necessary to apply other control inputs to eliminate the influence of the impulse on the system. When the impulse is used as the control input, it is necessary to apply the impulsive control to the system at the appropriate instant to achieve the control goal. In the 1950s, research on impulses mainly focused on impulsive perturbation. In 1999, Yang (1999) proposed the impulsive control theory and obtained sufficient conditions for realizing the asymptotic stability of the control system.

Since the twenty-first century, impulsive control methods have been widely used in chaotic synchronization, complex networks, neural networks, and fractional-order systems. In addition, scholars have also taken various factors affecting the stability of the system into consideration in the control system model studied, such as random disturbance, actuator saturation, time-delay phenomenon, and so on. In addition, impulsive control has also been applied in practice, such as satellite orbit transfer and formation maintenance, interest rate adjustment in financial markets, ecosystem management and regulation, system security, confidential communications, and the current sharing control of DC microgrid system.

In recent years, impulsive control has been widely used in the problem of achieving consensus of MASs. Liu et al. (2019), Wang et al. (2022), Chen et al. (2022), and You et al. (2023) realized the impulsive consensus of time-delayed MASs. Jiang et al. (2012), Zhang et al. (2017), Amiri Mehra et al. (2022), and Hu and Mu (2023) studied the impulsive consensus of MASs under switching topology. Tan et al. (2019), Li et al. (2020), Xie et al. (2022), and Han et al. (2023) combined impulsive control and event-triggered/self-triggered control to achieve the consensus of MASs. Guan et al. (2012, 2015), Ding et al. (2013), and Zhou et al. (2022) designed a distributed impulsive controller for second-order MASs to achieve consensus. Wang and Yang (2017), Ma et al. (2018), Bohner et al. (2022), and Zhao et al. (2022) realized the impulsive consensus of fractional MASs. The above literature only apply the control input at the impulsive instants, while the Wu et al. (2012), Liu et al. (2018b), Zhai et al. (2020), Chen et al. (2021), Lu and Li (2022), and Donganont and Liu (2023) exert the control input at both the continuous process and the impulsive instant. In this case, the impulsive control only plays an auxiliary role to make the error system converge faster. Although continuous and impulsive hybrid control can make the convergence speed faster, this hybrid control method still needs to continuously measure the state or error information of the system, which cannot save the information transmission loss and cannot fully utilize the advantages of impulsive control. Therefore, this book only considers the control scheme that applies the control input at the impulsive instants.

Based on the advantages of impulsive protocol, many researchers have investigated the formation problem of MASs via distributed impulsive control. Wang et al. (2014) designed an impulsive protocol with and without the input delay for the formation based on sampled information and just utilized information about relative position. Liu et al. (2014) firstly applied the impulsive control protocol in MASs to achieve the formation control objective. Wang et al. (2020) investigated bipartite formation of second-order nonlinear MASs with hybrid impulses, state-feedback controllers, and impulsive protocols designed for second-order MASs based on the definition of bipartite formation. Wu et al. (2020) addressed multi-group formation tracking (MGFT) control problems for second-order nonlinear MASs with multiple leaders via distributed impulsive control methods. Han and Zeng (2022) investigated a leader-based formation control problem of MASs via asynchronous impulsive protocols with saturated feedback. Chu et al. (2023) considered the output formation-containment problem for MASs under multipoint multipattern false data injection attacks.

1.4 Preliminaries

This section lists the mathematical symbols and some basic lemmas often involved in this book.

1.4.1 Symbol Description

$\mathbb{R}$, $\mathbb{R}^n$, $\mathbb{R}^{n\times m}$ denote the real numbers, the n-dimensional Euclidean space, and the set of all $n \times m$ real matrices, respectively. Matrices, if not explicitly stated, are assumed to have compatible dimensions. The superscript 'T' stands for the transpose for a matrix. Let $\mathbb{N}_+ = \{1, 2, \ldots\}$. I_n is the n dimensional identity matrix. $\mathrm{diag}(d_1, \ldots, d_N)$ indicates the diagonal matrix with diagonal elements d_1 to d_N. $\otimes$ denotes the Kronecker product. $\lambda_{\max}(A)$ denotes the maximal eigenvalue of A.

1.4.2 Graph Theory

A weighted graph $\mathcal{G} = (\mathcal{V}, \mathcal{E}, \mathcal{A})$ presents the relationship between agents in MASs. $\mathcal{V} = \{1, \ldots, N\}$ and $\mathcal{E} = (\mathcal{V} \times \mathcal{V})$ denote N followers and edges sets. The adjacency matrix $\mathcal{A} = [a_{ij}] \in \mathbb{R}^{N\times N}$. If $(j, i) \in \mathcal{E}$, node j is a neighbor of i, and $a_{ij} \neq 0$; otherwise $a_{ij} = 0$. Denote neighbors set of i as $N_i = \{j | (j, i) \in \mathcal{E}\}$. Assume self-loops do not exist in $\mathcal{G}$. The Laplacian matrix $\mathcal{L} = [l_{ij}] \in \mathbb{R}^{N\times N}$ is $l_{ij} = \sum_{k=1,k\neq i}^{N} a_{ik}$, if $i = j$, otherwise $l_{ij} = -a_{ij}$.

Denote $\bar{\mathcal{G}} = (\bar{\mathcal{V}}, \bar{\mathcal{E}}, \bar{\mathcal{A}}_s)$, where $\bar{\mathcal{V}} = \{0, 1, \ldots, N\}$, $\bar{\mathcal{E}} = (\bar{\mathcal{V}} \times \bar{\mathcal{V}})$, and $\bar{\mathcal{A}}_s = [\bar{a}_{ij}] \in \mathbb{R}^{(N+1)\times(N+1)}$. Denote $C = \mathrm{diag}\{c_i\} \in \mathbb{R}^{N\times N}$ as pinning matrix, $c_i > 0$ means node i has communication with leader, otherwise $c_i = 0$.

Assumption 1.1 The graph $\bar{\mathcal{G}}$ has a directed spanning tree with leader being root.

Remark 1.1 Assumption 1.1 is very mild, and it includes almost all possible graph topology, such as graph $(\bar{\mathcal{G}})$ contains a spanning tree and at least the root node can get access to the leader node, graph $(\bar{\mathcal{G}})$ is disconnected and each separated subgroup is either a single node or contains a spanning tree.

Lemma 1.1 (Chen and Lewis 2011) *If the graph $\bar{\mathcal{G}}$ contains a directed spanning tree and the root node is the leader node, then* $\mathrm{Rank}(L + C) = N$.

1.4.3 System Stability Theory

Consider the state equation of the system as

$$\begin{cases} \dot{x} = f(x, t), t \in [t_0, \infty) \\ x(t_0) = x_0, \end{cases} \tag{1.1}$$

where $x \in \mathbb{R}^n$ is the state of system, $f(x, t) \in \mathbb{R}^n$ is a N dimensional vector function.

Definition 1.1 For every $t \geq 0$, if the state x_e satisfies

$$\dot{x}_e = f(x_e, t) = 0, \tag{1.2}$$

then x_e can be referred to as the equilibrium state of system (1.1).

Definition 1.2 For system (1.1), if the equilibrium state x_e stability in the Lyapunov sense, and when $t \to \infty$,

$$\lim_{t\to\infty} \|x(t) - x_e\| = 0, \tag{1.3}$$

then the equilibrium state x_e of the system is said to be asymptotically stable.

Definition 1.3 If x_e satisfies the stability in the Lyapunov sense, and for any initial state,

$$\lim_{t\to\infty} \|x(t) - x_e\| = 0, \tag{1.4}$$

then the equilibrium state x_e of the system is said to be global asymptotically stable.

For convenience, the rest of this book only considers the case where the equilibrium point is at the origin. For cases where the equilibrium point is not at the origin, the method of transferring the equilibrium point to the origin through translation can be used.

1.4.4 Impulsive Control Theory

Definition 1.4 The model of impulsive system: Assuming system P is a continuous system, $x(t) \in \mathbb{R}^n$ is the state of system. When the system is under impulsive effect, the value of system state $x(t)$ can change abruptly at the impulsive instant t_k, that is $x(t_k^+) = x(t_k) + J(k, x)$, $\{t_k \in \mathbb{R}^+, t_k < t_{k-1}, k \in \mathbb{N}^+\}$ denotes the set of impulsive instants.

The impulsive control systems are mainly divided into two classes, of which the expression of the first type of impulsive control system is shown as follows:

$$\begin{cases} \dot{x}(t) = f(t, x), \ t \neq t_k, \\ \Delta x(t_k) = J(k, x), \ t = t_k, \end{cases} \tag{1.5}$$

where $\Delta x(t_k) = x(t_k^+) - x(t_k^-)$ is the jump of state at the impulsive instant. It can be seen that this kind of system has no control effect in the continuous process $t \neq t_k$, so it can achieve the control goal only by applying the control at the discrete time $t = t_k$.

The expression of the second type of impulsive control system is

$$\begin{cases} \dot{x}(t) = f(t, x, \tilde{u}), \ t \neq t_k, \\ \Delta x(t_k) = J(k, x), \ t = t_k, \end{cases} \tag{1.6}$$

where $\tilde{u}$ is the continuous control input. Different from the first type of impulsive control system, the second type of impulsive control system still has a control input in the continuous process, that is, the control objective cannot be achieved only through the control input at the impulsive instant, and control $\tilde{u}$ needs to be applied in the continuous process for control system P.

In this book, in order to avoid the loss of system information transmission in the continuous process, only the first type of impulsive control system is considered in the subsequent chapters, that is, information transmission and control input are only carried out at the impulsive instant.

1.4.5 Relevant Lemmas

Assumption 1.2 Assume the continuous nonlinear function $f(\cdot) : \mathbb{R}^n \to \mathbb{R}^n$ satisfies the Lipschitz condition,

$$\|f(x_1) - f(x_2)\| \leq l \, \|x_1 - x_2\| \,, \tag{1.7}$$

where $l > 0$ is the Lipschitz constant.

Definition 1.5 For the matrix $A = [a_{ij}] \in \mathbb{R}^{m\times n}$, $B = [b_{ij}] \in \mathbb{R}^{p\times q}$, define the Kronecker product of matrix A and B as

$$A \otimes B = \begin{bmatrix} a_{11}B & a_{12}B & \cdots & a_{1n}B \\ a_{21}B & a_{22}B & \cdots & a_{2n}B \\ \vdots & \vdots & \ddots & \vdots \\ a_{m1}B & a_{m2}B & \cdots & a_{mn}B \end{bmatrix} \in \mathbb{R}^{mp\times nq} \tag{1.8}$$

Lemma 1.2 (Brewer 1978) *For the matrix $A = [a_{ij}] \in \mathbb{R}^{m\times n}$, $B = [b_{ij}] \in \mathbb{R}^{p\times q}$, $C = [c_{ij}] \in \mathbb{R}^{n\times k}$, and $D = [d_{ij}] \in \mathbb{R}^{q\times r}$ satisfy the following properties:*
(1) $(A \otimes B)(C \otimes D) = (AC) \otimes (BD)$;
(2) $(aA) \otimes B = A \otimes (aB)$;
(3) $(A + B) \otimes C = A \otimes C + B \otimes C$;
(4) $(A \otimes B)^T = A^T \otimes B^T$;
(5) If matrix A and matrix B are invertible, then $(A \otimes B)^{-1} = A^{-1} \otimes B^{-1}$.

Lemma 1.3 (Cao et al. 2005) *For two real matrixes W_1 and W_2, there exist a symmetrical matrix P and a constant $\varepsilon > 0$ that satisfy*

$$W_1^T W_2 + W_1 W_2^T \leq \varepsilon W_1^T P W_1 + \varepsilon^{-1} W_2^T P^{-1} W_2. \tag{1.9}$$

Lemma 1.4 *If the Lyapunov function $V(t)$ satisfies*

$$\dot{V}(t) \leq -\alpha V(t), \tag{1.10}$$

where the constant $\alpha > 0$, then the Lyapunov function $V(t)$ can converge to zero exponentially when $t \to \infty$, that is, it is satisfied

$$V(t) \leq V(0)e^{-\alpha t}. \tag{1.11}$$

Lemma 1.5 (Boyd et al. 1994) *For any linear matrix inequality $S = \begin{bmatrix} S_{11} & S_{12} \\ S_{12}^T & S_{22} \end{bmatrix} < 0$, the following inequality holds*
(1) $S_{22} < 0$;
(2) $S_{11} - S_{12} S_{22}^{-1} S_{22}^T < 0$.

Lemma 1.6 *For the positive definite A, if the maximum eigenvalue of A is $\lambda_{\max}$, and the minimum eigenvalue of A is $\lambda_{\min}$, then for arbitrary nonzero vector x, the following inequality is satisfied:*

$$\lambda_{\max} \geq \frac{x^T A x}{x^T x} \geq \lambda_{\min}. \tag{1.12}$$

Definition 1.6 The Dirac delta function $\delta^*(t)$ satisfies $\delta^*(t) = 0$ if $t \neq 0$, and $\int_{-\infty}^{+\infty} \delta^*(t)dt = 1$. In addition, the Dirac functions have the following property: Suppose the function $f(t)$ is a bounded function in time domain and is continuous at $t = t_0$, then

$$\int_{-\infty}^{+\infty} \delta^*(t - t_0) f(t) dt = f(t_0). \tag{1.13}$$

1.5 Organization

This book mainly studies the coordination problem of nonlinear MASs by impulsive control protocol. The problems considered in this book include the consensus and formation problem of MASs. The types of systems considered include integer-order systems and fractional-order systems. It also considers the problem of MASs being unable to precisely apply impulsive control at a fixed time in practice, as well as the inability to control all agents at the same impulsive instant. In addition, this book also considers the effects of stochastic disturbances, time delays, uncertain parameters, switching topologies, control gain error, and cyber attacks on the system, making the distributed impulsive control strategy designed in this book more practical.

This chapter mainly expounds the research background and significance of this book, then introduce the development history and research status of MASs and impulsive control and gives the relevant mathematical symbol definitions and lemmas in this book. Finally, the main content of this book is briefly introduced.

In Chap. 2, the consensus of nonlinear MASs with variable impulsive control protocol is investigated. Compared with the impulsive protocol with fixed impulsive instants, the variable impulsive consensus method is more reliable in practical. Firstly, the impulsive time window and odd impulsive control are considered. Secondly, a more flexible variable impulsive control protocol is proposed. In addition, the variable impulsive protocol is extended to intermittent impulsive protocol, which allows for larger impulsive control intervals. Based on the Lyapunov stability theory and impulsive differential equation, several sufficient conditions are obtained to realize the consensus of MASs. Finally, some simulations are presented to validate the effectiveness of variable and intermittent control methods.

In Chap. 3, for the situation that the impulsive input cannot be applied to all agents in MASs simultaneously at an impulsive instant, the asynchronous impulsive consensus control problem of MASs is studied. Firstly, the design of asynchronous impulsive controller is completed, that is, the consensus control objective can be achieved by controlling a part of agents. Then, based on impulsive differential equations and the Lyapunov stability theory, sufficient condition for asynchronous impulsive consensus control is obtained. Finally, we propose two schemes to reduce the conservatism of the asynchronous impulsive control method, thereby enhancing the practicability of the control method. Finally, two simulations are given to validate the correctness and effectiveness of the asynchronous impulsive protocol.

In Chap. 4, the stochastic consensus of delayed nonlinear MASs via impulsive control is investigated. Based on the stochastic system theory, impulsive differential equations, the Lyapunov stability theory, and the contradiction method, a sufficient condition for consensus of MASs subject to time delay and stochastic perturbation is obtained. Simulation results verify the effectiveness of the proposed distributed impulsive consensus protocol in this chapter.

In Chap. 5, the adaptive impulsive consensus for uncertain MASs under stochastic perturbations is studied. Firstly, a discontinuous adaptive law is proposed for a linear system, which only requires the transmission of state information at impulsive instants, thus reducing the communication cost and making the control system more practical. Then, based on the theories of the Lyapunov stability and impulsive differential equations, and the related properties of infinite series, the conditions of impulsive control gain and impulsive interval under discontinuous adaptive impulsive control law are given, which ensure the consensus of uncertain MASs under stochastic perturbation can be realized. Finally, two simulations are presented to verify the correctness and effectiveness of the discontinuous adaptive impulsive control protocol in this chapter.

In Chap. 6, the consensus of fractional-order nonlinear MASs via distributed impulsive protocol with control gain error is studied. Based on the theory of impulsive differential equations, the Lyapunov stability theory, and the Mittag–Leffler function, some sufficient conditions for achieving the consensus of fractional-order nonlinear

MASs are derived. It is worth mentioning that the influence of control gain error is considered, which is more rigorous and practical in real systems. Finally, some simulation results verify the correctness and effectiveness of the distributed impulsive control protocol with control gain error.

In Chap. 7, the impulsive consensus of MASs under cyber attacks is studied. Firstly, based on the recovery mechanism of DoS attacks, the control process is divided into three different windows, namely normal window, attack window, and regulate window. Then, based on the theories of impulsive differential equations and the Lyapunov stability, sufficient conditions are obtained to realize the consensus of MASs under DoS attacks. Then based on the theories of the Lyapunov stability, impulsive differential equation, and stochastic system, some sufficient conditions are derived to achieve the consensus of MASs under deception attacks where the mean-squared consensus errors are bounded, and the expression of the error bound is obtained simultaneously. In addition, based on the obtained sufficient conditions and the expression of the error bound, criteria for selecting control parameters, such as lower and upper bound of impulsive interval, impulsive control gain, and convergence rate are discussed to obtain better control performance. Finally, some necessary simulations are performed to validate the effectiveness of theoretical results in this chapter.

In Chap. 8, we discuss the formation of second-order MASs with switching topologies via time-delayed impulsive control. The key issue is how to design an appropriate control method to maintain a desired geometric formation between neighboring agents with input delays and switching topologies. First, the distributed impulsive protocol is designed for the follower agents. Then, by using the stability and discrete-time systems theory, conditions are derived to ensure the formation consensus. Finally, the numerical examples illustrate the effectiveness and correctness of the designed impulsive control algorithm.

The last chapters summarizes the main work of this book and gives some possible development directions of this research in theoretical research and practical application. Finally, future work is discussed.

References

Amiri Mehra AH, Shafieirad M, Zamani I (2022) Leader-following consensus considering effects of agents on each other via impulsive control: A topology dependent average dwell time approach. J Franklin Inst 359(16):8639–8668

Bohner M, Hristova S, Malinowska AB, Morgado ML, Almeida R (2022) A generalized proportional Caputo fractional model of multi-agent linear dynamic systems via impulsive control protocol. Commun Nonlinear Sci Numer Simul 115:106756

Boyd S, Cgaoui LE, Feron E, Balakrishnan V (1994) Linear matrix inequalities in system and control theory Philadelphia. SIAM, PA

Brewer JW (1978) Kronecker products and matrix calculus in system theory. IEEE Trans Circuits Syst 25(9):772–781

Cao J, Li H, Ho D (2005) Synchronization criteria of Lur'e systems with time-delay feedback control. Chaos, Solitons Fractals 23(4):1285–1298

Chen G, Lewis FL (2011) Distributed adaptive tracking control for synchronization of unknown networked lagrangian systems. IEEE Trans Syst Man Cybern B Cybern 41(3):805–816

Chen F, Chen Z, Liu X, Xiang L, Yuan Z (2008) Decentralized formation control of mobile agents: a unified framework. Physica A 387(19–20):4917–4926

Chen T, Peng S, Zhang Z (2021) Finite-time consensus of leader-following non-linear multi-agent systems via event-triggered impulsive control. IET Control Theory Appl 15(7):926–936

Chen B, Ding D, Tang Z (2022) Distributed impulsive consensus of nonlinear multi-agent systems via matrix measure protocol. In: Chinese control conference, (CCC), pp 4766–4771

Cheng H, Huang J (2022) Adaptive robust bearing-based formation control for multi-agent systems. Automatica 162:111509

Chu H, Gorbachev S, Yue D, Dou C (2023) Output formation containment for multiagent systems under multipoint multipattern FDI attacks: a resilient impulsive compensation control approach. IEEE Trans Cybern 54(4):2606–2617

Cong Y, Du H, Liu B, Zhang P, Li X (2023) Distributed constrained finite-time consensus algorithm for second-order multi-agent systems. Inf Sci 626:773–786

Ding L, Yu P, Liu ZW, Guan ZH, Feng G (2013) Consensus of second-order multi-agent systems via impulsive control using sampled hetero-information. Automatica 49(9):2881–2886

Do KD (2008) Output-feedback formation tracking control of unicycle-type mobile robots with limited sensing ranges. Robot Auton Syst 57(1):34–47

Do KD (2014) Practical formation control of multiple underactuated ships with limited sensing ranges. Robot Auton Syst 59(6):457–471

Dong G, Li H, Ma H, Lu J (2021) Finite-time consensus tracking neural network FTC of multi-agent systems. IEEE Trans Neural Netw Learn Syst 32(2):653–662

Donganont M, Liu X (2023) Scaled consensus problems of multi agent systems via impulsive protocols. Appl Math Model 116:532–546

Du H, Wen G, Wu D, Cheng Y, Lyu J (2020) Distributed fixed-time consensus for nonlinear heterogeneous multi-agent systems. Automatica 113:108797

Guan ZH, Liu ZW, Feng G, Jian M (2012) Impulsive consensus algorithms for second-order multi-agent networks with sampled information. Automatica 48(7):1397–1404

Guan ZH, Han GS, He DX, Feng G (2015) Impulsive multiconsensus of second-order multiagent networks using sampled position data. IEEE Trans Neural Netw Learn Syst 26(11):2678–2688

Guo X, Wei G, Ding D (2023) Fault-tolerant consensus control for discrete-time multi-agent systems: a distributed adaptive sliding-mode scheme. IEEE Trans Circuits Syst II Express Briefs 70(7):2515–2519

Han Y, Zeng Z (2022) Asynchronous impulsive protocols with asymmetric feedback saturation on leader-based formation control of multiagent systems. IEEE Trans Cybern 52(9):9931–9942

Han J, Zhang H, Wang Y, Ren H (2023) Output consensus problem for linear heterogeneous multi-agent systems with dynamic event-based impulsive control. IEEE Trans Syst Man Cybern: Syst 53(1):334–345

Hu Z, Mu X (2023) Impulsive consensus of stochastic multi-agent systems under semi-Markovian switching topologies and application. Automatica 150:110871

Hu W, Cheng Y, Yang C (2023) Leader-following consensus of linear multi-agent systems via reset control: a time-varying systems approach. Automatica 149:110824

Huang J, Wang W, Wen C, Zhou J, Li G (2020) Distributed adaptive leader-follower and leaderless consensus control of a class of strict-feedback nonlinear systems: a unified approach. Automatica 118:109021

Jiang B, Bi Q, Zheng S (2012) Impulsive consensus in directed networks of identical nonlinear oscillators with switching topologies. Commun Nonlinear Sci Numer Simul 17(1):378–387

Jiang Y, Liu L, Feng G (2024) Fully distributed adaptive control for output consensus of uncertain discrete-time linear multi-agent systems. Automatica 162:111531

Lee G, Chwa D (2017) Decentralized behavior-based formation control of multiple robots considering obstacle avoidance. Intel Serv Robot 11:127–138

Li G, Wang L (2023) Adaptive output consensus of heterogeneous nonlinear multi-agent systems: a distributed dynamic compensator approach. IEEE Trans Autom Control 68(4):2483–2489
Li W, Chen Z, Liu Z (2014) Formation control for nonlinear multi-agent systems by robust output regulation. Neurocomputing 140(22):114–120
Li X, Yang X, Cao J (2020) Event-triggered impulsive control for nonlinear delay systems. Automatica 117:108981
Li H, Xie L, Zhang X, Pan W (2023a) Distributed consensus control of nonlinear multi-agent systems with sensor uncertainty. Int J Robust Nonlinear Control 33(2):973–990
Li J, Yuan L, Chai T, Lewis FL (2023b) Consensus of nonlinear multiagent systems with uncertainties using reinforcement learning based sliding mode control. IEEE Trans Circuits Syst I Regul Pap 70(1):424–434
Liu X-K, He H, Wang Y-W, Xu Q, Guo F (2018a) Distributed hybrid secondary control for a DC microgrid via discrete-time interaction. IEEE Trans Energy Convers 33(4):1865–1875
Liu X, Zhang K, Xie WC (2018b) Consensus of multi-agent systems via hybrid impulsive protocols with time-delay. Nonlinear Anal Hybrid Syst 30:134–146
Liu X, Xiao JW, Chen D, Wang YW (2019) Dynamic consensus of nonlinear time-delay multi-agent systems with input saturation: an impulsive control algorithm. Nonlinear Dyn 97(2):1699–1710
Liu H, Ma T, Lewis FL, Wang Y (2020) Robust formation control for multiple quadrotors with nonlinearities and disturbances. IEEE Trans Cybern 50(4):1362–1371
Liu D, Liu Z, Chen CLP, Zhang Y (2021) Distributed adaptive fuzzy control approach for prescribed-time containment of uncertain nonlinear multi-agent systems with unknown hysteresis. Nonlinear Dyn 105(1):257–275
Liu M, Wang YW, Xiao JW, Liu XK (2014) Formation tracking control for multi-agent systems with nonlinear dynamics via impulsive control. In: The 26th Chinese control and decision conference (CCDC), pp 3669–3674
Lu X, Jia Y (2023) Scaled event-triggered resilient consensus control of continuous-time multi-agent systems under byzantine agents. IEEE Trans Netw Sci Eng 10(2):1157–1174
Lu X, Li H (2022) Consensus of singular linear multiagent systems via hybrid control. IEEE Trans Control Netw Syst 9(2):647–656
Lu R, Wu J, Zhan X, Yan H (2024) Practical finite-time and fixed-time containment for second-order nonlinear multi-agent systems with IDAs and Markov switching topology. Neurocomputing 573:127180
Ma T, Li T, Cui B (2018) Coordination of fractional-order nonlinear multi-agent systems via distributed impulsive control. Int J Syst Sci 49(1):1–14
Madhevan B, Sreekumar M (2013) Tracking algorithm using leader flower approach for multi robots. Procedia Eng 64:1426–1435
Nair RR, Karki H, Shukla A, Behera L, Jamshidi M (2018) Fault-tolerant formation control of nonholonomic robots using fast adaptive gain nonsingular terminal sliding mode control. IEEE Syst J 13(1):1006–1017
Nascimento TP, Moreira AP, Conceicao AS (2013) Multi-robot nonlinear model predictive formation control: moving target and target absence. Robot Auton Syst 61(12):1502–1515
Olfati-Saber R, Murray R (2004) Consensus problems in networks of agents with switching topology and time-delays. IEEE Trans Autom Control 49(9):1520–1533
Pan Z, Wang D, Deng H, Li K (2019) A virtual spring method for the multi-robot path planning and formation control. Int J Control Autom Syst 17(5):1272–1282
Ren W, Beard RW (2005) Consensus seeking in multi-agent systems under dynamically changing interaction topologies. IEEE Trans Autom Control 50(5):655–661
Samoilenko AM (1995) Impulsive differential equations. World Scientific
Sinafar B, Kharrati H, Badamchizadeh MA, Baradarannia M (2020) Distributed adaptive switching control of uncertain switched affine multi-agent systems. Nonlinear Anal Hybrid Syst 38:100941
Tan X, Cao J, Li X (2019) Consensus of leader-following multiagent systems: a distributed event-triggered impulsive control strategy. IEEE Trans Cybern 49(3):792–801

Wang F, Yang Y (2017) Leader-following exponential consensus of fractional order nonlinear multi-agents system with hybrid time-varying delay: A heterogeneous impulsive method. Physica A 482:158–172

Wang YW, Liu M, Liu ZW, Yi JW (2014) Formation tracking of the second-order multi-agent systems using position-only information via impulsive control with input delays. Appl Math Comput 246:572–585

Wang W, Huang C, Huang C, Cao J, Lu J, Wang L (2020) Bipartite formation problem of second-order nonlinear multi-agent systems with hybrid impulses. Appl Math Comput 370:124926

Wang Q, Dong X, Yu J, Lyu J, Ren Z (2021) Predefined finite-time output containment of nonlinear multi-agent systems with leaders of unknown inputs. IEEE Trans Circuits Syst I: Regular Papers. 68(8):3436–3448

Wang KP, Ding D, Tang Z, Feng J (2022) Leader-following consensus of nonlinear multi-agent systems with hybrid delays: distributed impulsive pinning strategy. Appl Math Comput 424:127031

Wang G, Liu M, Wang F, Chen Y (2023) A novel and elliptical lattice design of flocking control for multi-agent ground vehicles. IEEE Control Syst Lett 7:1159–1164

Wang Z, Jin Z, Li H (2024) Semi-global asymptotic state agreement of nonlinear multi-agent systems with communication delays under directed switching topologies. Nonlinear Anal Hybrid Syst 52:101458

Wu Q, Zhou J, Xiang L (2012) Impulsive consensus seeking in directed networks of multi-agent systems with communication time delays. Int J Syst Sci 43(8):1479–1491

Wu Z, Liu X, Sun J, Wang X (2020) Multi-group formation tracking control via impulsive strategy. Neurocomputing 411:487–497

Wu M, Wu L-B, Wang P (2024) Event-triggered adaptive leaderless consensus control for nonlinear multi-agent systems with unknown dead-zones and output constraints. Appl Math Comput 470:128585

Xiao F, Wang L, Chen J, Gao Y (2009) Finite-time formation control for multi-agent systems. Automatica 45(11):2605–2611

Xiao J, Yuan G, He J, Fang K, Wang Z (2023) Graph attention mechanism based reinforcement learning for multi-agent flocking control in communication-restricted environment. Inf Sci 620:142–157

Xie X, Wei T, Li X (2022) Hybrid event-triggered approach for quasi-consensus of uncertain multi-agent systems with impulsive protocols. IEEE Trans Circuits Syst I Regul Pap 69(2):872–883

Yang T (1999) Impulsive Control. IEEE Trans Autom Control 44(5):1081–1083

Yao D, Dou C, Yue D, Zhao N, Zhang T (2020) Adaptive neural network consensus tracking control for uncertain multi-agent systems with predefined accuracy. Nonlinear Dyn 101(4):2249–2262

Yao D, Li H, Lu R, Shi Y (2022) Event-based distributed sliding mode formation control of multi-agent systems and its applications to robot manipulators. Inf Sci 614:87–103

You L, Jiang X, Li B, Zhang X, Yan H, Huang T (2023) Control for nonlinear fuzzy time-delay multi-agent systems: two kinds of distributed saturation-constraint impulsive approach. IEEE Trans Fuzzy Syst 31(8):2861–2670

Yu J, Yu S (2020) An improved fixed-time bipartite flocking protocol for nonlinear multi-agent systems. Int J Control 95(4):900–905

Yu J, Dong X, Li Q, Ren Z (2018) Robust H-infinity guaranteed cost time-varying formation tracking for high-order multiagent systems with time-varying delays. IEEE Trans Syst Man Cybern: Syst 50(4):1465–1475

Yu Z, Yu S, Jiang H, Hu C (2021) Distributed consensus for multi-agent systems via adaptive sliding mode control. Int J Robust Nonlinear Control 31(15):7125–7151

Zhai X, Wen G, Peng Z, Zhang X (2020) Leaderless and leader-following fixed-time consensus for multiagent systems via impulsive control. Int J Robust Nonlinear Control 30(13):5253–5266

Zhang H, Ji H, Ye Z, Senping T, Zhang H, Li C (2017) Impulsive consensus of multi-agent systems with stochastically switching topologies. Nonlinear Anal Hybrid Syst 26:212–224

Zhang C, Wang J, Zhang D, Shao W (2019) Synchronization and tracking of multi-spacecraft formation attitude control using adaptive sliding mode. Asian J Control 21(2):832–846
Zhao D, Li Y, Li S, Gao Z, Zhang C (2022) Distributed event-triggered impulsive tracking control for fractional-order multiagent networks. IEEE Trans Syst Man Cybern: Syst 52(7):4544–4556
Zhou P, Chen BM (2024) Distributed optimal solutions for multiagent pursuit-evasion games for capture and formation control. IEEE Trans Industr Electron 71(5):5224–5234
Zhou B, Yang Y, Li L, Xu X, Yang R (2022) Bipartite consensus for second-order multi-agent systems with external disturbances via constraint data-sampled impulsive method. Commun Nonlinear Sci Numer Simul 109:106314

Chapter 2
Distributed Variable Impulsive Consensus of Multi-agent Systems

2.1 Introduction

In practical systems, it is hard to input an impulse exactly at an exact fixed moment because of the physical and environmental constraints. The impulsive time window method allows the impulsive control to be exerted in a time range instead of requiring a specific moment (Li et al. 2017).

Due to the flexibility of control methods based on impulsive time windows compared to impulsive control with fixed impulsive instants, there have been many research results on impulsive time window. Feng et al. (2015), Wang et al. (2016a, b), and Zou et al. (2018) achieved asymptotic stability of a system by using impulsive time window method. Chen et al. (2015) and Wu et al. (2020) studied the stability and synchronization issues of memristive chaotic circuit systems under impulsive time window, respectively. Wang et al. (2015, 2016c) and Zhou et al. (2017, 2018) investigated the synchronization problem of neural networks based on impulsive time window. Xu et al. (2020), Long et al. (2021), Wang et al. (2022), and Wei et al. (2023) realized the consensus of MASs by using impulsive time window protocol. Specifically, Xu et al. (2020) achieved the consensus of MASs with impulsive time window via a hybrid protocol. Long et al. (2021) studied the finite-time consensus of MASs through impulsive time window protocol. Wang et al. (2022) realized the quasi-consensus of stochastic systems based on impulsive time window method. Wei et al. (2023) achieved the average consensus of MASs by using impulsive time window and intermittent impulsive control.

The impulsive time window method can relieve the shortage of fixed impulsive protocol, because the impulsive instant can be chosen within a region (i.e., the impulsive time window). In addition, the odd impulsive control and its extension form have the same function. That is to say, the consensus can be realized when the impulsive sequence $\{t_1, t_3, t_5, \ldots\}$ or $\{t_{2k-1}\}$ $k = 1, 2, \ldots$ is determined, and there is not any restriction for the even impulsive sequence $\{t_2, t_4, t_6, \ldots\}$ or $\{t_{2k}\}$. In this case, we can choose impulsive instant t_{2k} within (t_{2k-1}, t_{2k+1}).

Z. Zhang et al., *Distributed Impulsive Coordination of Multi-Agent Systems*, Intelligent Control and Learning Systems 15, https://doi.org/10.1007/978-981-97-6262-0_2

Although the impulsive time window and odd impulsive method has some flexibility, it still needs to determine the left and right endpoints that can be applied at each impulsive instant. Moreover, the existing impulsive time window method still limits the range of impulse to a small window, and each small window only contains an impulse. In the odd impulsive control protocol, it is needed to fix the control instants of all odd impulses, which means that only the even impulsive instants are variable. Therefore, how to design a more flexible and convenient variable impulsive controller is an important problem that needs to be solved. In order to increase the flexibility and practicality of impulsive control, the variable impulsive control considered in this chapter allows multiple impulses to be exerted in each time window, and the impulses can be applied at any instant in the time window without fixing the instant of any impulse. In addition, in order to allow the impulsive controller to relax for a period of time, this chapter extends the variable impulsive control to intermittent impulsive control.

2.2 Problem Description

Consider a system with N agents is described by

$$\dot{x}_i(t) = Ax_i(t) + f(x_i(t)) + u_i(t), \tag{2.1}$$

where $x_i(t) \in \mathbb{R}^n$ is the state of agent i, $(i = 1, 2, \ldots N)$, $f : \mathbb{R}^n \to \mathbb{R}^n$ is a continuous nonlinear function of agent i. $A \in \mathbb{R}^{n\times n}$ is a known matrix, and $x_i = [x_{i1}, x_{i2}, \ldots, x_{in}]^T$.

Consider the leader node as

$$\dot{x}_0(t) = Ax_0(t) + f(x_0(t)), \tag{2.2}$$

where $x_0(t) \in \mathbb{R}^n$ is the state of leader node.

Definition 2.1 The global synchronization error is defined as

$$\delta_i(t) = x_i(t) - x_0(t). \tag{2.3}$$

Definition 2.2 The local neighborhood synchronization error for agent i is defined as

$$e_i(t) = \sum_{j\in N_i} a_{ij}(x_i(t) - x_j(t)) + c_i(x_i(t) - x_0(t)). \tag{2.4}$$

The distributed impulsive controller for agent i is designed as follows:

$$u_i(t) = \sum_{k=1}^{\infty} b_k \left(\sum_{j \in N_i} a_{ij}(x_i(t) - x_j(t)) + c_i(x_i(t) - x_0(t)) \right) \delta^*(t - t_k), \tag{2.5}$$

where $i = 1, 2, \ldots, N$, $k \in \mathbb{N}_+$, b_k is the impulsive control gain. The Dirac delta function $\delta^*(t)$ satisfies $\delta^*(t) = 0$ if $t \neq 0$, and $\int_{-\infty}^{+\infty} \delta^*(t)dt = 1$. In this chapter, $x_i(t)$ satisfies left continuous at t_k, i.e., $x_i(t_k) = x_i(t_k^-)$.

Then the distributed impulsive controlled system is obtained as

$$\begin{cases} \dot{x}_i(t) = Ax_i(t) + f(x_i(t)), \quad t \neq t_k, \\ \Delta x_i(t_k) = x_i(t_k^+) - x_i(t_k^-) = b_k e_i(t_k) \\ \qquad = b_k(\sum_{j \in N_i} a_{ij}(x_i(t_k) - x_j(t_k)) + c_i(x_i(t_k) - x_0(t_k))), \\ x_i(t_0^+) = x_i(t_0), \quad t_0 \geq 0, \end{cases} \tag{2.6}$$

where $\Delta x_i(t_k)$ is the jump of the state of node i at impulsive instant t_k.

From (2.1), (2.2), and (2.5), the error system is obtained as

$$\begin{cases} \dot{\delta}_i(t) = A\delta_i(t) + f(x_i(t)) - f(x_0(t)), t \neq t_k, \\ \Delta\delta_i(t_k) = b_k e_i(t_k). \end{cases} \tag{2.7}$$

By using the Kronecker product, the system (2.7) can be rewritten as

$$\begin{cases} \dot{\delta}(t) = (I_N \otimes A)\delta(t) + \bar{f}(x(t), \bar{x}_0(t)), \quad t \neq t_k, \\ \Delta\delta(t_k) = b_k((L + C) \otimes I_n)\delta(t_k), \end{cases} \tag{2.8}$$

where $x(t) = [x_1^T(t), x_2^T(t), \ldots, x_n^T(t)]^T \in \mathbb{R}^{Nn}$, $\bar{x}_0(t) = 1_N \otimes x_0(t) \in \mathbb{R}^{Nn}$, $\bar{f}(x(t), \bar{x}_0(t)) = [f^T(x_1(t)) - f^T(x_0(t)), \ldots, f^T(x_N(t)) - f^T(x_0(t))]^T \in \mathbb{R}^{Nn}$.

The aim of the consensus problem is to design the impulsive controller for all nodes i in $\mathcal{G}$, such that all nodes synchronize to the leader node for any initial conditions, i.e.,

$$\lim_{t \to \infty} \delta(t) = 0. \tag{2.9}$$

For simplicity, in the rest of this chapter, all variables will be denoted without the time parameter t, i.e., $x := x(t)$, $x_i := x_i(t)$, $\delta := \delta(t)$, $\delta_i := \delta_i(t)$.

2.3 Main Results

2.3.1 Impulsive Time Window Protocol

In this section, the main results will be given to accomplish consensus of MASs with impulsive time window. The radiuses and centers of impulsive time window satisfy the following Assumption 2.1, and the relation of some instants is shown in Fig. 2.1. The shaded area in Fig. 2.1 indicates the possible range of actual impulsive instants, which corresponds to the so-called impulsive time window.

Assumption 2.1

$$\tau_{k-1}^l < t_{k-1} < \tau_{k-1}^r < \tau_k^l < t_k < \tau_k^r < \tau_{k+1}^l < t_{k+1} < \tau_{k+1}^r, \quad k \in \mathbb{N}+$$

where $\tau_k^l = \tau_k - r_k$ and $\tau_k^r = \tau_k + r_k$ are the left and right endpoints of the kth impulsive time window, respectively. τ_k and r_k are the center and radius of the kth time window.

Theorem 2.1 Suppose that Assumptions 1.1, 1.2, and 2.1 hold, if there exists a constant $\xi > 1$ such that

$$(\lambda_A + 2l)(\Delta_{k+1} - \mu_{k+1}) + \ln(\lambda_k \xi) \leq 0, \tag{2.10}$$

where the parameters λ_A and λ_k are the maximum eigenvalue of $A + A^T$ and $(b_k(L + C) + I_N)^T(b_k(L + C) + I_N)$, respectively, and $0 < \Delta_k = \tau_k - \tau_{k-1} < \infty$, $0 < \mu_k = r_k - r_{k-1} < \infty$ are the impulsive centers distance and radius difference, respectively. Then the consensus of MASs is realized.

Proof Choose the Lyapunov function as $V(t) = \delta^T\delta$, for $t \in (t_{k-1}, t_k]$,

$$\begin{aligned} D^+V(t) &= \dot{\delta}^T\delta + \delta^T\dot{\delta} \\ &= \delta^T(I_N \otimes (A + A^T))\delta + 2\delta^T\bar{\psi}(x, \bar{x}_0) \\ &\leq (\lambda_A + 2l)\delta^T\delta \\ &= (\lambda_A + 2l)V(t), \end{aligned} \tag{2.11}$$

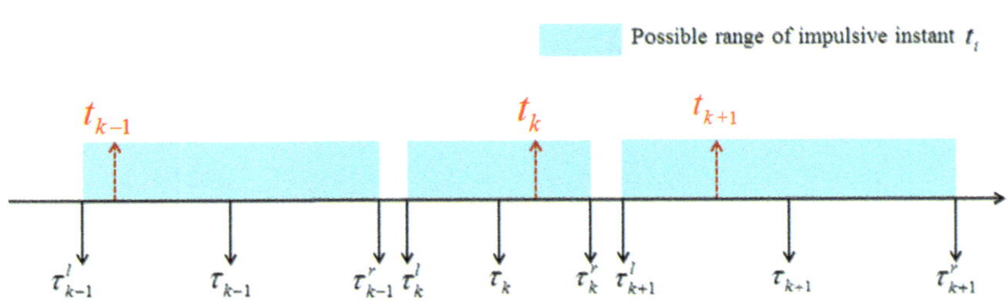

Fig. 2.1 The diagram of impulsive time window

which leads to

$$V(t) \leq V(t_{k-1}^{+}) \exp((\lambda_A + 2l)(t - t_{k-1})). \tag{2.12}$$

When $t = t_k$, one can get

$$\begin{aligned} & V(t_k^{+}) \\ &= \delta^T(t_k^{+})\delta(t_k^{+}) \\ &= ((b_k((L+C)\otimes I_n) + I_{nN})\,\delta(t_k))^T\,(b_k((L+C)\otimes I_n) + I_{nN})\,\delta(t_k) \\ &= \delta^T(t_k)\left(\left((b_k(L+C) + I_N)^T b_k((L+C) + I_N)\right)\otimes I_n\right)\delta(t_k) \\ &\leq \lambda_k V(t_k). \end{aligned} \tag{2.13}$$

For $t \in (t_0, \tau_1 - r_1]$, it follows from (2.12) that

$$V(t) \leq V(t_0) \exp((\lambda_A + 2l)(t - t_0)). \tag{2.14}$$

Similarly, if $t \in (\tau_1 - r_1, t_1]$, from (2.12), one can derive

$$V(t) \leq V(t_0) \exp((\lambda_A + 2l)(t - t_0)). \tag{2.15}$$

When $t \in (t_1, \tau_2 - r_2]$, from (2.12) and (2.13), it yields

$$\begin{aligned} V(t) &\leq V(t_1^{+}) \exp((\lambda_A + 2l)(t - t_1)) \\ &\leq \lambda_1 V(t_1) \exp((\lambda_A + 2l)(t - t_1)) \\ &\leq \lambda_1 V(t_0) \exp((\lambda_A + 2l)(t_1 - t_0)) \exp((\lambda_A + 2l)(t - t_1)) \\ &= \lambda_1 V(t_0) \exp((\lambda_A + 2l)(t - t_0)). \end{aligned} \tag{2.16}$$

Therefore, for $t \in (\tau_1 - r_1, \tau_2 - r_2]$, one can get

$$V(t) \leq \lambda_1^{\kappa_1} V(t_0) \exp((\lambda_A + 2l)(t - t_0)), \tag{2.17}$$

where $\kappa_k = \begin{cases} 0, t < t_k \\ 1, t \geq t_k \end{cases}, k \in \mathbb{N}_+$.

In general, for $t \in (\tau_k - r_k, \tau_{k+1} - r_{k+1}]$, one can obtain

$$V(t) \leq V(t_0)\lambda_1\lambda_2\cdots\lambda_{k-1}\lambda_k^{\kappa_k} \exp((\lambda_A + 2l)(t - t_0)). \tag{2.18}$$

From condition (2.10), one can get

$$\lambda_k \exp(\lambda_A + 2l)(\Delta_{k+1} - \mu_{k+1}) \leq \frac{1}{\xi}. \tag{2.19}$$

Thus, for $t \in (\tau_k - r_k, \tau_{k+1} - r_{k+1}]$, it has

$$
\begin{aligned}
V(t) &\leq V(t_0)\lambda_1\lambda_2\cdots\lambda_{k-1}\lambda_k^{\kappa_k}\exp((\lambda_A+2l)(t-t_0))\\
&\leq V(t_0)\lambda_1\lambda_2\cdots\lambda_{k-1}\lambda_k^{\kappa_k}\exp((\lambda_A+2l)(\tau_{k+1}-r_{k+1}-t_0))\\
&\leq V(t_0)\exp((\lambda_A+2l)(\tau_1-r_1-t_0))(\lambda_1\exp((\lambda_A+2l)(\Delta_2-\mu_2)))\cdots\\
&\quad(\lambda_{k-1}\exp((\lambda_A+2l)(\Delta_k-\mu_k)))\lambda_k^{\kappa_k}\exp((\lambda_A+2l)(\Delta_{k+1}-\mu_{k+1}))\\
&\leq \frac{1}{\xi^{k-1}}V(t_0)\lambda_k^{\kappa_k}\exp((\lambda_A+2l)(\Delta_{k+1}-\mu_{k+1}+\tau_1-r_1-t_0)).
\end{aligned}
\tag{2.20}
$$

Since $V(t_0)\lambda_k^{\kappa_k}\exp((\lambda_A+2l)(\Delta_{k+1}-\mu_{k+1}+\tau_1-r_1-t_0))$ is a finite constant, and $1/\xi^{k-1}\to 0$ as $k\to\infty$. Thus the consensus error δ can globally asymptotically converge to zero. The proof is completed.

Remark 2.1 Let $\tau_k^l=\tau_k-r_k$ be the left endpoint of kth impulsive time window. Then the condition (2.10) in Theorem 2.1 can be changed as

$$
(\lambda_A+2l)(\tau_{k+1}^l-\tau_k^l)+\ln(\lambda_k\xi)\leq 0, \tag{2.21}
$$

where $\tau_{k+1}^l-\tau_k^l=\tau_{k+1}-r_{k+1}-(\tau_k-r_k)=\Delta_{k+1}-\mu_{k+1}$. The equivalence of conditions (2.10) and (2.21) means that as long as the distance of left endpoint of adjacent impulsive time window satisfied condition (2.21), the consensus of MASs can be realized.

Remark 2.2 The consensus condition (2.10) can be transformed into the following form:

$$
\frac{1}{\exp((\lambda_A+2l)(\Delta_{k+1}-\mu_{k+1}))\lambda_k}\geq\xi.
$$

That is to say, the consensus condition is that the constraint combination $\frac{1}{\exp((\lambda_A+2l)(\Delta_{k+1}-\mu_{k+1}))\lambda_k}$, which includes system parameters λ_A+2l, impulsive control gain b_k, impulsive interval $\Delta_{k+1}-\mu_{k+1}=\tau_{k+1}^l-\tau_k^l$, and the matrix $L+C$, is greater than a finite constant $\xi>1$. Furthermore, it is easy to conclude that smaller ξ can obtain larger acceptable interval $\Delta_{k+1}-\mu_{k+1}$, which shows the existence and necessity of ξ in the consensus condition.

Remark 2.3 In the process of proving Theorem 2.1, it is required that $\lambda_k\in(0,1)$ in (2.13). This is because to ensure that the application of impulsive control can reduce the global consensus error $\delta(t)$ of MASs, it is necessary to make the value of the Lyapunov function $\delta^T(t)\delta(t)$ smaller after impulsive instants than before impulsive instants. In fact, if $\lambda_k\geq 1$, it means that after an impulsive control, the global consensus error of MASs may not decrease or even increase. Then the application of impulsive control not only wastes control resources, but may even have a counterproductive effect. Therefore, in this chapter, parameter $\lambda_k\in(0,1)$ is a necessary condition for achieving impulsive consensus of MASs. From the definition of λ_k, one gets λ_k is the maximum eigenvalue of $(b_k(L+C)+I_N)^T(b_k(L+C)+I_N)$, so $\lambda_k\in(0,1)$ can be guaranteed by selecting a suitable impulsive control gain b_k. In addition, from (2.10) and (2.21), one gets if λ_k is smaller, the obtained distance between the left endpoints of the adjacent impulsive time window is larger, which means that the selectable range of impulsive instant is larger. Therefore, to ensure the maximum selectable range of impulsive intervals, the appropriate parameter b_k should be selected to minimize λ_k.

In the proof of Theorem 2.1, we discuss the interval $t \in (\tau_k - r_k, \tau_{k+1} - r_{k+1}] = (\tau_k^l, \tau_{k+1}^l]$, i.e., the interval between two adjacent left endpoints. If the interval is changed into $t \in (\tau_{k-1}, \tau_k]$, similarly, i.e., two centers distance of the adjacent impulsive time windows, then we can obtain the following Theorem 2.2.

Theorem 2.2 Suppose that Assumptions 1.1, 1.2, and 2.1 hold, if there exists a constant $\xi > 1$ such that

$$(\lambda_A + 2l)(\tau_{k+1} - \tau_k) + \ln(\lambda_k \xi) \leq 0, \tag{2.22}$$

where λ_A and λ_k have same definitions with Theorem 2.1. Then the consensus of MASs is achieved.

Proof Choose the Lyapunov function as $V(t) = \delta^T \delta$, similar to the proof of Theorem 2.1, i.e., (2.11) and (2.12), for $t \in (t_{k-1}, t_k]$, it yields

$$V(t) \leq V(t_{k-1}^+) \exp((\lambda_A + 2l)(t - t_{k-1})). \tag{2.23}$$

When $t = t_k$, similar to the proof of Theorem 2.1, i.e., (2.13), one can get

$$V(t_k^+) \leq \lambda_k V(t_k). \tag{2.24}$$

For $t \in [t_0, \tau_1]$, there are three cases (see Table 2.1 and Figs. 2.2, 2.3, and 2.4) to be considered

Case 1. It follows from (2.23) that

$$V(t) \leq V(t_0) \exp((\lambda_A + 2l)(t - t_0)). \tag{2.25}$$

Case 2. It follows from (2.23) and (2.24) that

$$\begin{aligned} V(t) &\leq V(t_1^+) \exp((\lambda_A + 2l)(t - t_1)) \\ &\leq \lambda_1 V(t_1) \exp((\lambda_A + 2l)(t - t_1)) \\ &\leq \lambda_1 V(t_0) \exp((\lambda_A + 2l)(t_1 - t_0)) \exp((\lambda_A + 2l)(t - t_1)) \\ &= \lambda_1 V(t_0) \exp((\lambda_A + 2l)(t - t_0)). \end{aligned} \tag{2.26}$$

Table 2.1 The possible cases for $t \in [t_0, \tau_1]$

$t \leq \tau_1$	$t \in [t_0, t_1]$	Case 1
	$t \in (t_1, \tau_1]$	Case 2
$t > \tau_1$	$t \in [t_0, \tau_1]$	Case 3

Fig. 2.2 The diagram of Case 1 for $t \in [t_0, \tau_1]$

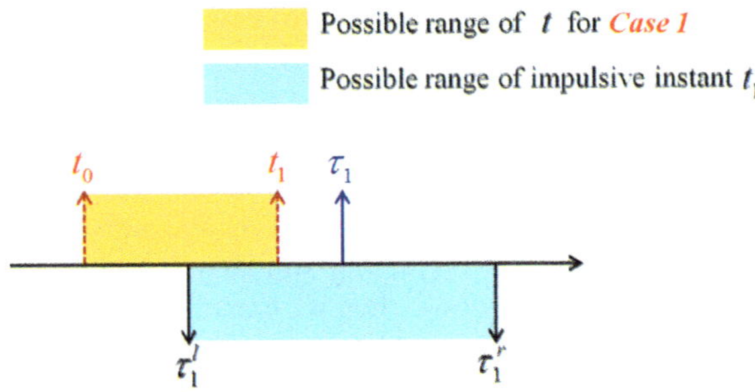

Fig. 2.3 The diagram of Case 2 for $t \in [t_0, \tau_1]$

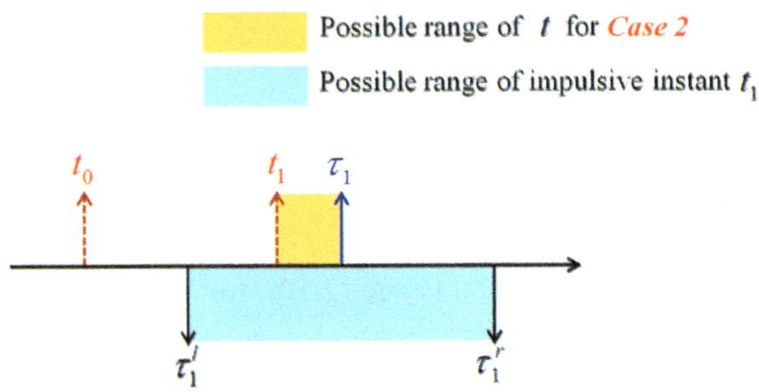

Fig. 2.4 The diagram of Case 3 for $t \in [t_0, \tau_1]$

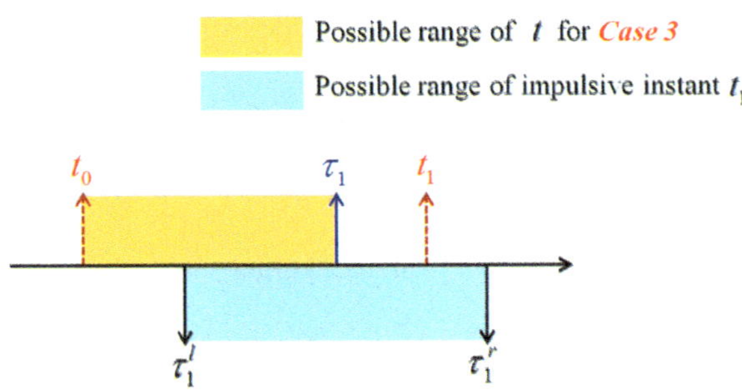

Case 3. It follows from (2.23) that

$$V(t) \leq V(t_0)\exp((\lambda_A + 2l)(t - t_0)). \tag{2.27}$$

Therefore, from (2.25) to (2.27), for $t \in (t_0, \tau_1]$, one gets

$$V(t) \leq \lambda_1^{\kappa_1} V(t_0)\exp((\lambda_A + 2l)(t - t_0)), \tag{2.28}$$

where κ_k has the same definition with Theorem 2.1.

For $t \in (\tau_1, \tau_2]$, there are eight cases (see Table 2.2 and Figs. 2.5, 2.6, 2.7, 2.8, 2.9, 2.10, 2.11, and 2.12) to be considered.

Table 2.2 The possible cases for $t \in (\tau_1, \tau_2]$

$t_1 \le \tau_1$	$t_2 \le \tau_2$	$t \in (\tau_1, t_2]$	Case 1
		$t \in (t_2, \tau_2]$	Case 2
	$t_2 > \tau_2$	$t \in (\tau_1, \tau_2]$	Case 3
$t_1 > \tau_1$	$t_2 \le \tau_2$	$t \in (\tau_1, t_1]$	Case 4
		$t \in (t_1, t_2]$	Case 5
		$t \in (t_2, \tau_2]$	Case 6
	$t_2 > \tau_2$	$t \in (\tau_1, t_1]$	Case 7
		$t \in (t_1, \tau_2]$	Case 8

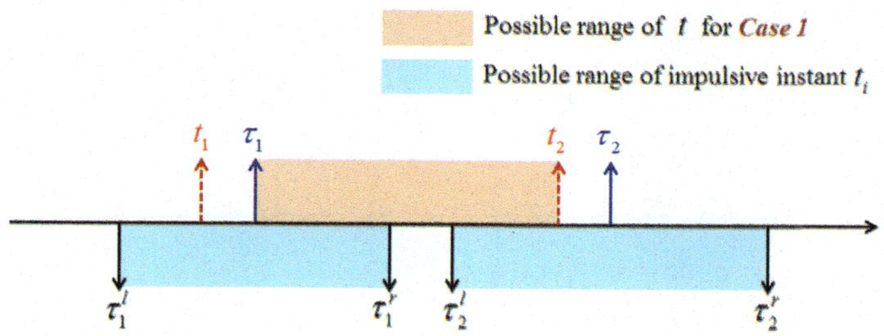

Fig. 2.5 The diagram of Case 1 for $t \in (\tau_1, \tau_2]$

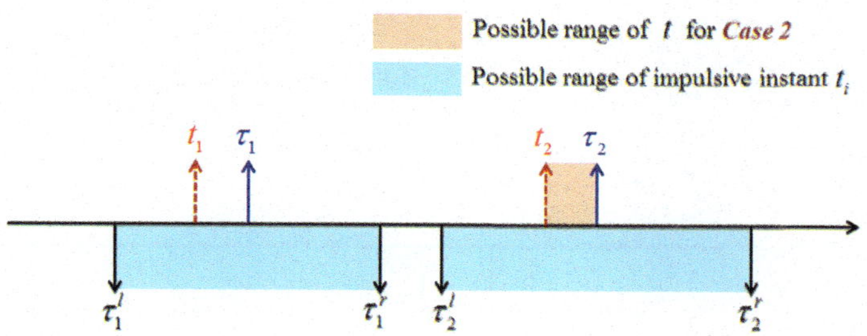

Fig. 2.6 The diagram of Case 2 for $t \in (\tau_1, \tau_2]$

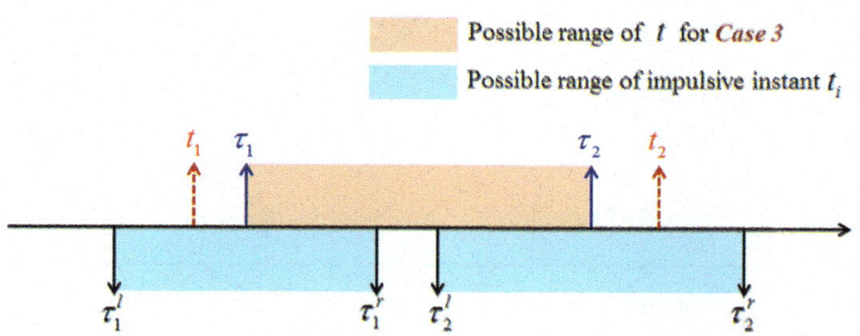

Fig. 2.7 The diagram of Case 3 for $t \in (\tau_1, \tau_2]$

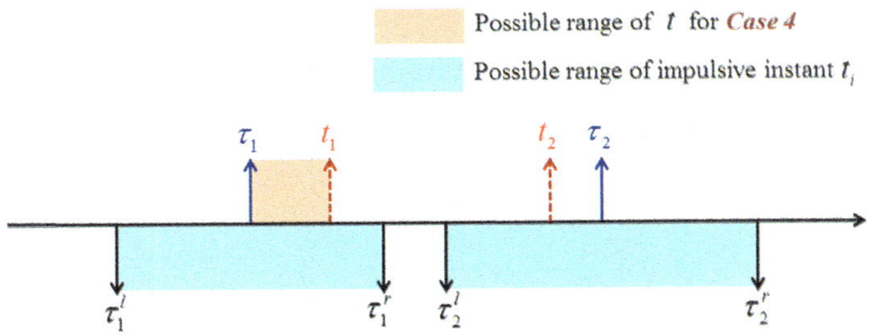

Fig. 2.8 The diagram of Case 4 for $t \in (\tau_1, \tau_2]$

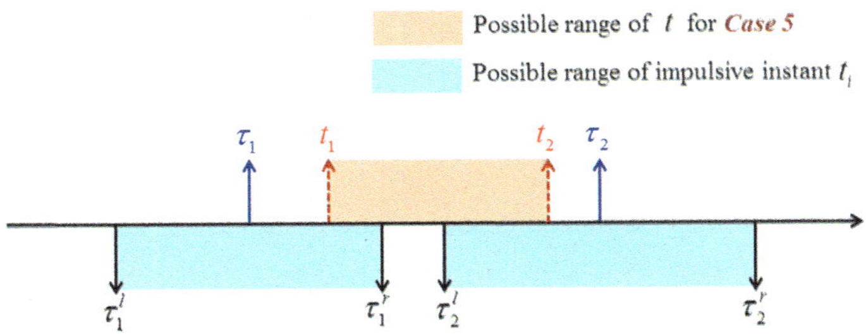

Fig. 2.9 The diagram of Case 5 for $t \in (\tau_1, \tau_2]$

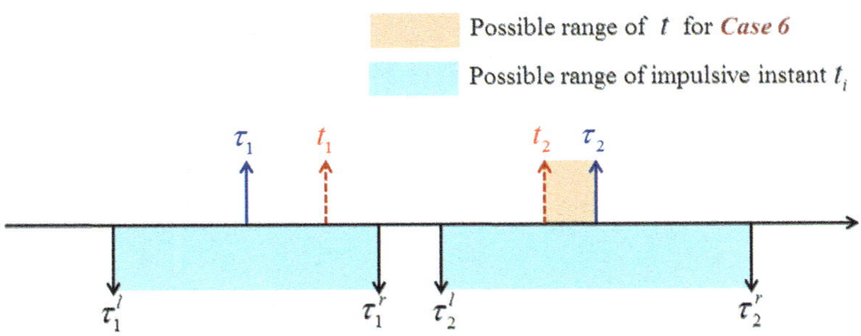

Fig. 2.10 The diagram of Case 6 for $t \in (\tau_1, \tau_2]$

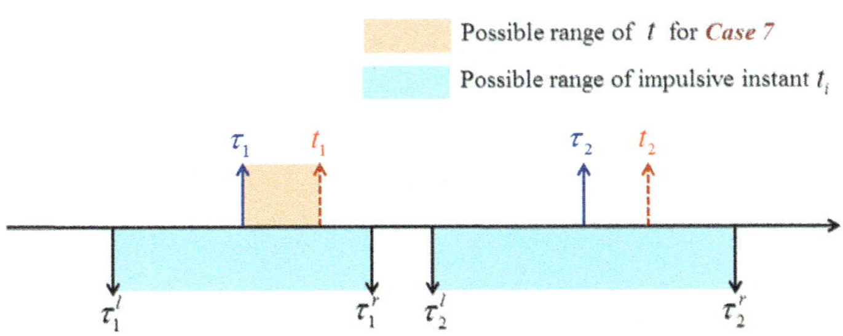

Fig. 2.11 The diagram of Case 7 for $t \in (\tau_1, \tau_2]$

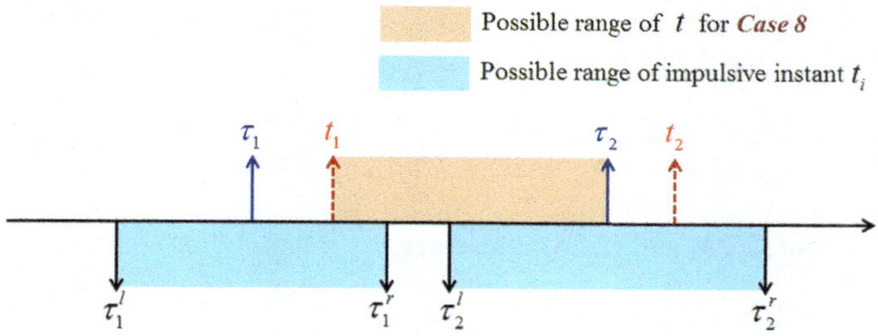

Fig. 2.12 The diagram of Case 8 for $t \in (\tau_1, \tau_2]$

Case 1. It follows from (2.23) and (2.24) that

$$\begin{aligned} V(t) &\le V(t_1^+)\exp((\lambda_A + 2l)(t - t_1)) \\ &\le \lambda_1 V(t_1)\exp((\lambda_A + 2l)(t - t_1)) \\ &\le \lambda_1 V(t_0)\exp((\lambda_A + 2l)(t_1 - t_0))\exp((\lambda_A + 2l)(t - t_1)) \\ &= \lambda_1 V(t_0)\exp((\lambda_A + 2l)(t - t_0)). \end{aligned} \tag{2.29}$$

Case 2. It follows from (2.23) and (2.24) that

$$\begin{aligned} V(t) &\le V(t_2^+)\exp((\lambda_A + 2l)(t - t_2)) \\ &\le \lambda_2 V(t_2)\exp((\lambda_A + 2l)(t - t_2)) \\ &\le \lambda_2\lambda_1 V(t_0)\exp((\lambda_A + 2l)(t_2 - t_0))\exp((\lambda_A + 2l)(t - t_2)) \\ &= \lambda_1\lambda_2 V(t_0)\exp((\lambda_A + 2l)(t - t_0)). \end{aligned} \tag{2.30}$$

Case 3. From (2.23) and (2.24), one gets

$$\begin{aligned} V(t) &\le V(t_1^+)\exp((\lambda_A + 2l)(t - t_1)) \\ &\le \lambda_1 V(t_1)\exp((\lambda_A + 2l)(t - t_1)) \\ &\le \lambda_1 V(t_0)\exp((\lambda_A + 2l)(t_1 - t_0))\exp((\lambda_A + 2l)(t - t_1)) \\ &= \lambda_1 V(t_0)\exp((\lambda_A + 2l)(t - t_0)). \end{aligned} \tag{2.31}$$

Case 4. It follows from (2.23) that

$$V(t) \le V(t_0)\exp((\lambda_A + 2l)(t - t_0)). \tag{2.32}$$

Case 5. It follows from (2.23) and (2.24) that

$$\begin{aligned} V(t) &\le V(t_1^+)\exp((\lambda_A+2l)(t-t_1)) \\ &\le \lambda_1 V(t_1)\exp((\lambda_A+2l)(t-t_1)) \\ &\le \lambda_1 V(t_0)\exp((\lambda_A+2l)(t_1-t_0))\exp((\lambda_A+2l)(t-t_1)) \\ &= \lambda_1 V(t_0)\exp((\lambda_A+2l)(t-t_0)). \end{aligned} \tag{2.33}$$

Case 6. From (2.23) and (2.24), one can get

$$\begin{aligned} V(t) &\le V(t_2^+)\exp((\lambda_A+2l)(t-t_2)) \\ &\le \lambda_2 V(t_2)\exp((\lambda_A+2l)(t-t_2)) \\ &\le \lambda_2\lambda_1 V(t_0)\exp((\lambda_A+2l)(t_2-t_0))\exp((\lambda_A+2l)(t-t_2)) \\ &= \lambda_1\lambda_2 V(t_0)\exp((\lambda_A+2l)(t-t_0)). \end{aligned} \tag{2.34}$$

Case 7. It follows from (2.23) that

$$V(t) \le V(t_0)\exp((\lambda_A+2l)(t-t_0)). \tag{2.35}$$

Case 8. From (2.23) and (2.24), one can get

$$\begin{aligned} V(t) &\le V(t_1^+)\exp((\lambda_A+2l)(t-t_1)) \\ &\le \lambda_1 V(t_1)\exp((\lambda_A+2l)(t-t_1)) \\ &\le \lambda_1 V(t_0)\exp((\lambda_A+2l)(t_1-t_0))\exp((\lambda_A+2l)(t-t_1)) \\ &= \lambda_1 V(t_0)\exp((\lambda_A+2l)(t-t_0)). \end{aligned} \tag{2.36}$$

Based on (2.29) to (2.36), for $t \in (\tau_1, \tau_2]$, one gets

$$V(t) \le \lambda_1^{\kappa_1}\lambda_2^{\kappa_2} V(t_0)\exp((\lambda_A+2l)(t-t_0)). \tag{2.37}$$

In general, for $t \in (\tau_{k-1}, \tau_k]$, one can derive

$$V(t) \le V(t_0)\lambda_1\lambda_2\cdots\lambda_{k-2}\lambda_{k-1}^{\kappa_{k-1}}\lambda_k^{\kappa_k}\exp((\lambda_A+2l)(t-t_0)). \tag{2.38}$$

From (2.22), one gets

$$\lambda_k \exp(\lambda_A+2l)(\tau_{k+1}-\tau_k) < \frac{1}{\xi}. \tag{2.39}$$

Thus, for $t \in (\tau_{k-1}, \tau_k]$,

$$
\begin{aligned}
&V(t) \\
&\leq V(t_0)\lambda_1\lambda_2\cdots\lambda_{k-2}\lambda_{k-1}^{\kappa_{k-1}}\lambda_k^{\kappa_k}\exp((\lambda_A+2l)(t-t_0)) \\
&\leq V(t_0)\lambda_1\lambda_2\cdots\lambda_{k-2}\lambda_{k-1}^{\kappa_{k-1}}\lambda_k^{\kappa_k}\exp((\lambda_A+2l)(\tau_k-t_0)) \\
&\leq V(t_0)\exp((\lambda_A+2l)((\tau_1-t_0))\lambda_1\exp((\lambda_A+2l)(\tau_2-\tau_1))\cdots \\
&\quad\lambda_{k-2}\exp((\lambda_A+2l)(\tau_{k-1}-\tau_{k-2}))\lambda_{k-1}^{\kappa_{k-1}}\exp((\lambda_A+2l)(\tau_k-\tau_{k-1}))\lambda_k^{\kappa_k} \\
&\leq \frac{1}{\xi^{k-2}}V(t_0)\lambda_k^{\kappa_k}\lambda_{k-1}^{\kappa_{k-1}}\exp((\lambda_A+2l)(\tau_k-\tau_{k-1})) \\
&\quad\exp((\lambda_A+2l)(\tau_1-t_0)).
\end{aligned}
\tag{2.40}
$$

Since $V(t_0)\lambda_k^{\kappa_k}\lambda_{k-1}^{\kappa_{k-1}}\exp((\lambda_A+2l)(\tau_k-\tau_{k-1}))\exp((\lambda_A+2l)(\tau_1-t_0))$ is a finite constant, and $1/\xi^{k-2}\to 0$ as $k\to\infty$. Thus the consensus error δ can globally asymptotically converge to zero. The proof is completed.

Remark 2.4 It follows from the condition (2.10) that both the center τ_k and the radius r_k are needed. In fact, from (2.10) and (2.21), only the left endpoints of the adjacent impulsive time windows are necessary. Therefore, in this case, the condition (2.21) is less conservative than (2.10) of Theorem 2.1. By comparison, it follows from the condition (2.22) of Theorem 2.2 that only the center τ_k of the adjacent impulsive time windows is necessary. The choice of condition (2.21) or (2.22) can depend on the practical system to be considered.

2.3.2 *Odd Impulsive Control Protocol*

In this section, we consider another variable impulsive protocol, odd impulsive sequence, which means that we only need to determine the instants of odd impulses t_{2k-1}, and the even impulsive instants t_{2k} can be chosen arbitrarily between adjacent odd impulsive instants. The diagram of odd impulsive protocol is presented in Fig. 2.13.

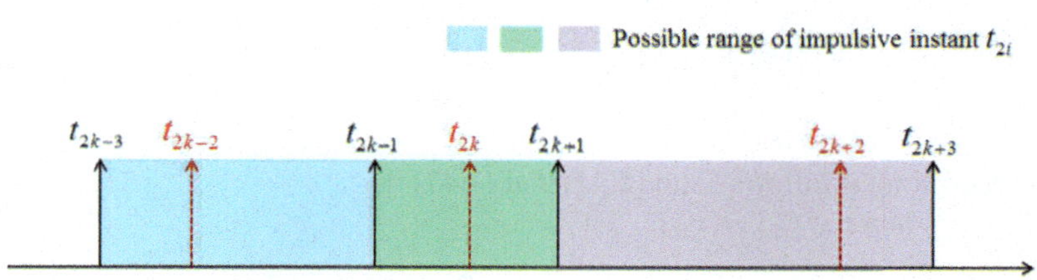

Fig. 2.13 The diagram of odd impulsive sequence

Theorem 2.3 Suppose that Assumptions 1.1 and 1.2 hold, if there exists a constant $\xi > 1$ such that

$$(\lambda_A + 2l)(t_{2k+1} - t_{2k-1}) + \ln(\lambda_{2k-1}\lambda_{2k}\xi) \leq 0, \tag{2.41}$$

where λ_A and λ_k have the same definitions with Theorems 2.1 and 2.2. Then the consensus of MASs can be realized.

Proof Choose the Lyapunov function as $V(t) = \delta^T\delta$, similar to Theorem 2.1, i.e., (2.11) and (2.12),when $t \in (t_{k-1}, t_k]$, one gets

$$V(t) \leq V(t_{k-1}^+)\exp((\lambda_A + 2l)(t - t_{k-1})). \tag{2.42}$$

When $t = t_k$, similar to Theorem 2.1, i.e., (2.13), one gets

$$V(t_k^+) \leq \lambda_k V(t_k). \tag{2.43}$$

For $t \in (t_0, t_1]$, it follows from (2.42) that

$$V(t) \leq V(t_0)\exp((\lambda_A + 2l)(t - t_0)) \tag{2.44}$$

and

$$V(t_1) \leq V(t_0)\exp((\lambda_A + 2l)(t_1 - t_0)). \tag{2.45}$$

For $t \in (t_1, t_2]$, it follows from (2.42), (2.43), and (2.45) that

$$\begin{aligned} V(t) &\leq V(t_1^+)\exp((\lambda_A + 2l)(t - t_1)) \\ &\leq \lambda_1 V(t_1)\exp((\lambda_A + 2l)(t - t_1)) \\ &\leq \lambda_1 V(t_0)\exp((\lambda_A + 2l)(t_1 - t_0))\exp((\lambda_A + 2l)(t - t_1)) \\ &= \lambda_1 V(t_0)\exp((\lambda_A + 2l)(t - t_0)). \end{aligned} \tag{2.46}$$

In general, for $t \in (t_k, t_{k+1}]$,

$$\begin{aligned} V(t) &\leq V(t_k^+)\exp((\lambda_A + 2l)(t - t_k)) \\ &\leq \lambda_1\lambda_2\ldots\lambda_k V(t_0)\exp((\lambda_A + 2l)(t - t_0)). \end{aligned} \tag{2.47}$$

Furthermore, it follows from (2.41) and (2.47) that

Case 1 When $t \in (t_{2k-1}, t_{2k}]$,

$$
\begin{aligned}
V(t) &\leq V(t_0) \prod_{i=1}^{2k-1} \lambda_i \exp((\lambda_A + 2l)(t - t_0)) \\
&\leq V(t_0) \prod_{i=1}^{2k-1} \lambda_i \exp((\lambda_A + 2l)(t_{2k} - t_0)) \\
&= V(t_0)\lambda_1\lambda_2 \exp((\lambda_A + 2l)(t_3 - t_1)) \cdots \\
&\quad \lambda_{2k-3}\lambda_{2k-2} \exp((\lambda_A + 2l)(t_{2k-1} - t_{2k-3})) \\
&\quad \lambda_{2k-1} \exp((\lambda_A + 2l)(t_{2k} - t_{2k-1})) \exp((\lambda_A + 2l)(t_1 - t_0)) \\
&\leq \frac{1}{\xi^{k-1}} V(t_0) \exp((\lambda_A + 2l)(t_1 - t_0)) \\
&\quad \lambda_{2k-1} \exp((\lambda_A + 2l)(t_{2k} - t_{2k-1})).
\end{aligned} \tag{2.48}
$$

Case 2 When $t \in (t_{2k}, t_{2k+1}]$

$$
\begin{aligned}
V(t) &\leq V(t_0) \prod_{i=1}^{2k} \lambda_i \exp((\lambda_A + 2l)(t - t_0)) \\
&\leq V(t_0) \prod_{i=1}^{2k} \lambda_i \exp((\lambda_A + 2l)(t_{2k+1} - t_0)) \\
&= V(t_0)\lambda_1\lambda_2 \exp((\lambda_A + 2l)(t_3 - t_1)) \cdots \\
&\quad \lambda_{2k-1}\lambda_{2k} \exp(((\lambda_A + 2l)(t_{2k+1} - t_{2k-1})) \\
&\quad \exp(\lambda_A + 2l)(t_1 - t_0)) \\
&\leq \frac{1}{\xi^{k}} V(t_0) \exp(\lambda_A + 2l)(t_1 - t_0)).
\end{aligned} \tag{2.49}
$$

Since $(\lambda_A + 2l)$, $(t_1 - t_0)$, and $(t_{2k} - t_{2k-1})$ are finite constants, and $1/\xi^k \to 0$ as $k \to \infty$. Thus, it follows from (2.48) and (2.49) that the consensus error δ can globally asymptotically converge to zero. The proof is completed.

If we extend the odd impulsive protocol in Theorem 2.3 to the more general form, the following corollary can be obtained.

Corollary 2.1 *Suppose that Assumptions 1.1 and 1.2 hold, if there exist a bounded integer $n_0 > 0$ and a constant $\xi > 1$ such that*

$$
(\lambda_A + 2l)(t_{kn_0+1} - t_{(k-1)n_0+1}) + \ln(\lambda_{(k-1)n_0+1} \cdots \lambda_{kn_0}\xi) \leq 0. \tag{2.50}
$$

Then the consensus of MASs is achieved.

Proof Similar to the proof of Theorem 2.3, for $t \in (t_k, t_{k+1}]$, one gets

$$
\begin{aligned}
V(t) &\leq V(t_k^+) \exp((\lambda_A + 2l)(t - t_k)) \\
&\leq \lambda_1\lambda_2 \ldots \lambda_k V(t_0) \exp((\lambda_A + 2l)(t - t_0)).
\end{aligned} \tag{2.51}
$$

Furthermore, it follows from (2.50) and (2.51) that

Case 1: When $t \in (t_{(k-1)n_0+1}, t_{(k-1)n_0+2}]$,

$$\begin{aligned}
& V(t) \\
& \leq V(t_0) \prod_{i=1}^{(k-1)n_0+1} \lambda_i \exp((\lambda_A + 2l)(t - t_0)) \\
& \leq V(t_0) \prod_{i=1}^{(k-1)n_0+1} \lambda_i \exp((\lambda_A + 2l)(t_{(k-1)n_0+2} - t_0)) \\
& = V(t_0)\lambda_1\lambda_2 \cdots \lambda_{n_0} \exp((\lambda_A + 2l)(t_{n_0+1} - t_1)) \\
& \quad \lambda_{n_0+1}\lambda_{n_0+2} \cdots \lambda_{2n_0} \exp((\lambda_A + 2l)(t_{2n_0+1} - t_{n_0+1})) \cdots \\
& \quad \lambda_{(k-2)n_0+1}\lambda_{(k-2)n_0+2} \cdots \lambda_{(k-1)n_0} \exp((\lambda_A + 2l)(t_{(k-1)n_0+1} - t_{(k-2)n_0+1})) \\
& \quad \lambda_{(k-1)n_0+1} \exp((\lambda_A + 2l)(t_{(k-1)n_0+2} - t_{(k-1)n_0+1})) \\
& \quad \exp((\lambda_A + 2l)(t_1 - t_0)) \\
& \leq \frac{1}{\xi^{k-1}} V(t_0)\lambda_{(k-1)n_0+1} \exp((\lambda_A + 2l)(t_{(k-1)n_0+2} - t_{(k-1)n_0+1})) \\
& \quad \exp((\lambda_A + 2l)(t_1 - t_0)).
\end{aligned} \tag{2.52}$$

Case 2: When $t \in (t_{(k-1)n_0+2}, t_{(k-1)n_0+3}]$,

$$\begin{aligned}
& V(t) \\
& \leq V(t_0) \prod_{i=1}^{(k-1)n_0+2} \lambda_i \exp((\lambda_A + 2l)(t - t_0)) \\
& \leq V(t_0) \prod_{i=1}^{(k-1)n_0+2} \lambda_i \exp((\lambda_A + 2l)(t_{(k-1)n_0+3} - t_0)) \\
& = V(t_0)\lambda_1\lambda_2 \cdots \lambda_{n_0} \exp((\lambda_A + 2l)(t_{n_0+1} - t_1)) \\
& \quad \lambda_{n_0+1}\lambda_{n_0+2} \cdots \lambda_{2n_0} \exp((\lambda_A + 2l)(t_{2n_0+1} - t_{n_0+1})) \cdots \\
& \quad \lambda_{(k-2)n_0+1}\lambda_{(k-2)n_0+2} \cdots \lambda_{(k-1)n_0} \exp((\lambda_A + 2l)(t_{(k-1)n_0+1} - t_{(k-2)n_0+1})) \\
& \quad \lambda_{(k-1)n_0+1}\lambda_{(k-1)n_0+2} \exp((\lambda_A + 2l)(t_{(k-1)n_0+3} - t_{(k-1)n_0+1})) \\
& \quad \exp((\lambda_A + 2l)(t_1 - t_0)) \\
& \leq \frac{1}{\xi^{k-1}} V(t_0)\lambda_{(k-1)n_0+1}\lambda_{(k-1)n_0+2} \exp((\lambda_A + 2l)(t_{(k-1)n_0+3} - t_{(k-1)n_0+1})) \\
& \quad \exp((\lambda_A + 2l)(t_1 - t_0)).
\end{aligned} \tag{2.53}$$

Case n_0: When $t \in (t_{kn_0}, t_{kn_0+1}]$,

$$
\begin{aligned}
&V(t)\\
&\le V(t_0)\prod_{i=1}^{kn_0}\lambda_i \exp((\lambda_A+2l)(t-t_0))\\
&\le V(t_0)\prod_{i=1}^{kn_0}\lambda_i \exp((\lambda_A+2l)(t_{kn_0+1}-t_0))\\
&= V(t_0)\lambda_1\lambda_2\cdots\lambda_{n_0}\exp((\lambda_A+2l)(t_{n_0+1}-t_1))\\
&\lambda_{n_0+1}\lambda_{n_0+2}\cdots\lambda_{2n_0}\exp((\lambda_A+2l)(t_{2n_0+1}-t_{n_0+1}))\cdots\\
&\lambda_{(k-2)n_0+1}\lambda_{(k-2)n_0+2}\cdots\lambda_{(k-1)n_0}\exp((\lambda_A+2l)(t_{(k-1)n_0+1}-t_{(k-2)n_0+1}))\\
&\lambda_{(k-1)n_0+1}\lambda_{(k-1)n_0+2}\cdots\lambda_{kn_0}\exp((\lambda_A+2l)(t_{kn_0+1}-t_{(k-1)n_0+1}))\\
&\exp((\lambda_A+2l)(t_1-t_0))\\
&\le \frac{1}{\xi^k}V(t_0)\exp((\lambda_A+2l)(t_1-t_0)).
\end{aligned}
\tag{2.54}
$$

Since $V(t_0)\lambda_k^{\kappa_k}\lambda_{k-1}^{\kappa_{k-1}}\exp((\lambda_A+2l)(\tau_k-\tau_{k-1}))\exp((\lambda_A+2l)(\tau_1-t_0))$ is a finite constant, and $1/\xi^{k-2}\to 0$ as $k\to\infty$. Thus the consensus error δ can globally asymptotically converge to zero. The proof is completed.

Remark 2.5 The condition (2.50) is the extension of (2.41). As the especial cases, $n_0=1$ is the situation of full impulsive sequence (correspond to the common fixed time impulsive control), $n_0=2$ is the situation of odd impulsive control. From (2.41) or (2.50), if $n_0=2$, we need to choose odd impulsive sequence $\{t_{2k-1}\}$ (i.e., $t_1, t_3, t_5, \ldots$), and the even impulsive instant t_{2k} can be chosen within the fixed range (i.e., (t_{2k-1}, t_{2k+1})). From (2.50), if $n_0=3$, we need to choose the impulsive sequence $\{t_{3k-2}\}$ (i.e., $t_1, t_4, t_7, \ldots$), and the impulsive instants t_{3k-1} (i.e., $t_2, t_5, t_8, \ldots$) and t_{3k} (i.e., $t_3, t_6, t_9, \ldots$) can be chosen within the fixed range (i.e., (t_{3k-2}, t_{3k+1})). By comparison, it concludes that larger n_0 is helpful to obtain larger possible impulsive control region.

2.3.3 A More Flexible Variable Impulsive Control Protocol

It should be noted that impulsive time window method still limits the range of impulse to a small window, and each small window only contains an impulse. Moreover, in the odd impulsive control protocol, it is need to fix the control instants of all odd impulses, which means that only the even impulsive instants are variable. Therefore, how to design a more flexible and convenient variable impulsive controller is an important problem that needs to be solved. In this section, to increase the practicability of impulsive control, a more flexible variable impulsive control protocol is proposed, each window contains multiple impulses, and the moments of impulses are selected arbitrarily.

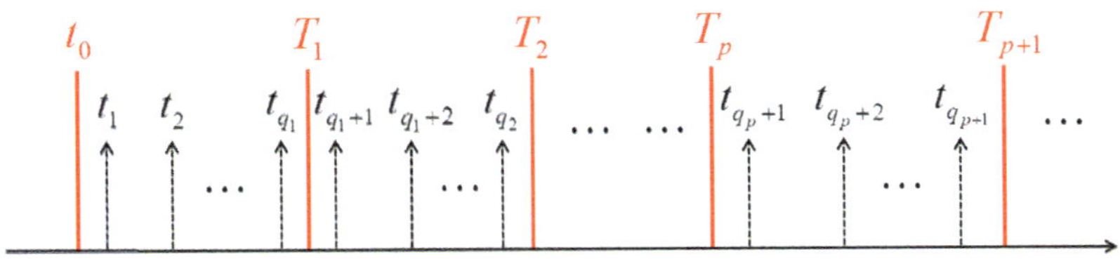

Fig. 2.14 Diagram of the more flexible variable impulsive control

Figure 2.14 presents the more flexible variable impulsive strategy. In Fig. 2.14, t_0 is the initial moment, t_{q_p} is the moment of the last impulse in pth window, T_p is the right endpoint of pth window, and q_p denotes the number of impulses before $t = T_p$. Let $q_0 = 0$, the number of impulses in pth window is $q_p - q_{p-1}$.

Theorem 2.4 Suppose the Assumptions 1.1 and 1.2 hold, if

$$(\lambda_A + 2l)(T_p - T_{p-1}) + \ln(\bar{\lambda}_p \xi) \leq 0, \tag{2.55}$$

where the constant $\xi > 1$, λ_A is the maximum eigenvalue of $A + A^T$, $\bar{\lambda}_p = \lambda_{q_{p-1}+1}\lambda_{q_{p-1}+2}\cdots\lambda_{q_p}$, $\lambda_k (k \in \mathbb{N}^+)$ has the same definition with Theorem 2.1 then the consensus of MASs can be realized.

Proof Choose the Lyapunov function as $V(t) = \delta^T\delta$, similar to Theorem 2.1, i.e., (2.11) and (2.12),for $t \in (t_{k-1}, t_k]$, one can obtain

$$V(t) \leq V(t_{k-1}^+)\exp((\lambda_A + 2l)(t - t_{k-1})). \tag{2.56}$$

When $t = t_k$, similar to Theorem 2.1, i.e., (2.13), one can get

$$V(t_k^+) \leq \lambda_k V(t_k). \tag{2.57}$$

When $t \in (t_0, t_1]$, from (2.56),

$$V(t) \leq V(t_0)\exp((\lambda_A + 2l)(t - t_0)), \tag{2.58}$$

which leads to

$$V(t_1) \leq V(t_0)\exp((\lambda_A + 2l)(t_1 - t_0)). \tag{2.59}$$

When $t \in (t_1, t_2]$, from (2.56), (2.57), and (2.59), one gets

$$\begin{aligned} V(t) &\leq V(t_1^+)\exp((\lambda_A + 2l)(t - t_1)) \\ &\leq \lambda_1 V(t_1)\exp((\lambda_A + 2l)(t - t_1)) \\ &\leq \lambda_1 V(t_9)\exp((\lambda_A + 2l)(t - t_0)). \end{aligned} \tag{2.60}$$

For arbitrary $t \in (t_{q_1}, t_{q_1+1}]$, one obtains

$$\begin{aligned}V(t) &\leq V(t_{q_1}^+)\exp((\lambda_A+2l)(t-t_{q_1}))\\ &\leq \lambda_1\exp((\lambda_A+2l)(t_2-t_1))\cdots\lambda_{q_1}\exp((\lambda_A+2l)(t-t_{q_1}))\\ &\quad V(t_0)\exp((\lambda_A+2l)(t_1-t_0))\\ &= V(t_0)\lambda_1\lambda_2\cdots\lambda_{q_1}\exp((\lambda_A+2l)(t-t_0)).\end{aligned} \tag{2.61}$$

Since $T_1 \in (t_{q_1}, t_{q_1+1}]$, from (2.61),

$$V(T_1) = V(t_0)\lambda_1\lambda_2\cdots\lambda_{q_1}\exp((\lambda_A+2l)(T_1-t_0)). \tag{2.62}$$

When $t = T_2$,

$$\begin{aligned}V(T_2) &\leq V(T_1)\lambda_{q_1+1}\lambda_{q_1+2}\cdots\lambda_{q_2}\exp((\lambda_A+2l)(T_2-T_1))\\ &\leq V(t_0)\lambda_1\lambda_2\cdots\lambda_{q_1}\exp((\lambda_A+2l)(T_1-t_0))\\ &\quad \lambda_{q_1+1}\lambda_{q_1+2}\cdots\lambda_{q_2}\exp((\lambda_A+2l)(T_2-T_1)).\end{aligned} \tag{2.63}$$

In general, when $T = T_p$,

$$\begin{aligned}V(T_p) &\leq V(T_{p-1})\lambda_{q_{p-1}+1}\lambda_{q_{p-1}+2}\cdots\lambda_{q_p}\exp((\lambda_A+2l)(T_p-T_{p-1}))\\ &\leq V(t_0)\lambda_1\lambda_2\cdots\lambda_{q_1}\exp((\lambda_A+2l)(T_1-t_0))\lambda_{q_1+1}\lambda_{q_1+2}\\ &\quad \cdots\lambda_{q_2}\exp((\lambda_A+2l)(T_2-T_1))\cdots\\ &\quad \lambda_{q_{p-1}+1}\lambda_{q_{p-1}+2}\cdots\lambda_{q_p}\exp((\lambda_A+2l)(T_p-T_{p-1}))\\ &= V(t_0)\bar{\lambda}_1\exp((\lambda_A+2l)(T_1-t_0))\bar{\lambda}_2\exp((\lambda_A+2l)(T_2-T_1))\cdots\\ &\quad \bar{\lambda}_p\exp((\lambda_A+2l)(T_p-T_{p-1})).\end{aligned} \tag{2.64}$$

From (2.55) in Theorem 2.4, one gets

$$V(t_p) \leq V(t_0)\left(\frac{1}{\xi}\right)^{p-1}. \tag{2.65}$$

Since $V(t_0)$ is a bounded constant, and $\lim_{p\to\infty} 1/\xi^p = 0$, then $\lim_{p\to\infty} V(T_p) = 0$.

When $t \in (T_p, t_{q_p+1}]$, from (2.56),

$$\begin{aligned}V(t) &\leq V(t_{q_p+1})\\ &\leq V(T_p)\exp((\lambda_A+2l)(t_{q_p+1}-T_p)).\end{aligned} \tag{2.66}$$

When $t \in (t_{q_p+1}, t_{q_p+2}]$, from (2.56), (2.57), and (2.66),

$$\begin{aligned}V(t) &\leq V(t_{q_p+2})\\ &\leq \lambda_{q_p+1}V(t_{q_p+1})\exp((\lambda_A+2l)(t_{q_p+2}-t_{q_p+1}))\\ &\leq \lambda_{q_p+1}V(T_p)\exp((\lambda_A+2l)(t_{q_p+2}-T_p)).\end{aligned} \tag{2.67}$$

In general, when $t \in (t_{q_{p+1}}, T_{p+1}]$,

$$\begin{aligned} V(t) &\leq V(T_{p+1}) \\ &\leq \lambda_{q_p+1}\lambda_{q_p+2}\cdots\lambda_{q_{p+1}} V(T_p)\exp((\lambda_A + 2l)(T_{p+1} - T_p)). \end{aligned} \tag{2.68}$$

Since λ_{q_p+1}, λ_{q_p+2}, ..., $\lambda_{q_{p+1}}$, $V(T_p)$, $\exp((\lambda_A + 2l)(t_{q_p+1} - T_p))$, $\exp((\lambda_A + 2l)(t_{q_p+2} - T_p))$, ..., $\exp((\lambda_A + 2l)(T_{p+1} - T_p))$ are bounded constants, and $\lim_{p\to\infty} V(T_p) = 0$, then $\lim_{t\to\infty} V(t) = 0$. Therefore, the consensus of MASs is realized. This completes the proof.

Remark 2.6 Theorem 2.4 shows the relationship between $\bar{\lambda}_p$ and the length of window $T_p - T_{p-1}$. To use Theorem 2.4, the following steps are given.
Step 1: Determine the end time of each window, that is, determine the instants of $T_p (p \in \mathbb{N}^+)$.
Step 2: From Theorem 2.4, the $\bar{\lambda}_p$ is calculated as

$$\bar{\lambda}_p \leq \exp(-(\lambda_A + 2l)(T_p - T_{p-1}))/\xi.$$

Step 3: According to the range of $\bar{\lambda}_p$, determine the impulsive control gain b_k and the number of impulses to be exerted in (T_{p-1}, T_p).
Step 4: In (T_{p-1}, T_p), exert the required impulsive control at arbitrary impulsive instants.

Remark 2.7 To use the impulsive time window control method based on the left endpoint in Theorem 2.1, it is necessary to calculate the distance between the left endpoints of two adjacent impulsive time windows $\tau_{k+1}^l - \tau_k^l$ firstly. Then set the instant of the right endpoint τ_k^r, and should note that $\tau_k^r < \tau_{k+1}^l$. Finally, based on the left and right endpoint instants of the impulsive time window, apply an impulse within the time window, which satisfies $t_k \in (\tau_k^l, \tau_k^r)$. In summary, using the control method in Theorem 2.1, it is necessary to determine the instants of the left and right endpoints of the impulsive time window before applying impulses within the time window. This not only increases the complexity of control strategies, but also limits the instants of impulses to a small window.

Remark 2.8 To use the impulsive time window control method based on the center endpoint in Theorem 2.2, it is necessary to determine the distance between the center points of adjacent time windows $\tau_{k+1} - \tau_k$ firstly. Then, based on the instant of the selected center point τ_k, determine the radius $r_k = \tau_k - \tau_k^l = \tau_k^r - \tau_k$, that is, determine the time at the left and right endpoints of the impulsive time window. When determining the radius r_k, it is necessary to ensure that Assumption 2.1 holds, i.e., $\tau_{k-1}^r < \tau_k^l < \tau_k^r < \tau_{k+1}^l$. Finally, similar to the control method based on the left endpoint of the impulsive time window, the impulsive instant is determined within the time window (τ_k^l, τ_k^r). It can be seen that when using the center point-based impulsive time window control method, it is necessary to determine the center point, left endpoint, and right endpoint of the impulsive time window firstly. Moreover, when selecting the radius of the impulsive time window, it is also limited by the instants of the left and right endpoints of the two time windows, which is more complicated than the control scheme based on the left endpoint impulsive time window. Similarly, adopting the variable impulsive control scheme based on the center point of the impulsive time window still requires limiting the instant of impulse within a small window.

Remark 2.9 According to Remarks 2.7 and 2.8, the impulsive time window method still needs to limit the impulsive instant to a small window and determine the left or center point of each time window. Compared with Theorems 2.1 and 2.2, the variable impulsive control scheme designed in this section does not require determining the center point, left endpoint, and right endpoint of the time window for each impulse, making the method simpler and more convenient to use. In addition, the variable impulsive method designed in this section does not need to limit the impulsive instant to a small window, so the range of impulsive instant selection in this section is larger, and the application of impulsive control is more flexible, further solving the problem that impulses cannot be exerted at a fixed time.

Remark 2.10 It should be noted that the odd impulsive control in Theorem 2.3 still requires the determination of the instants for odd impulses, and even the more general form in Corollary 2.1 still requires the determination of the control instants for each k impulses. Therefore, this control method cannot fully solve the problem that impulsive control cannot be exerted at a precise fixed time. Compared with Theorem 2.3 and Corollary 2.1, the variable impulsive controller designed in this section does not require any fixed application instants of impulse, making the protocol more flexible in practice.

2.3.4 Intermittent Impulsive Control Protocol

In this section, the variable impulsive protocol is extended to the intermittent impulsive approach. In the intermittent impulsive protocol, the impulses only need to be exerted in control window.

Figure 2.15 is the diagram of intermittent impulsive protocol. In Fig. 2.5, τ_1^*, τ_2^*, ..., τ_p^* are the switching instants from free window to control window, τ_1^r, τ_2^r, ..., τ_p^r are the right endpoints of the first, second, and pth windows, respectively. The following theorem is given to realize the bipartite consensus of MASs by using the intermittent impulsive control.

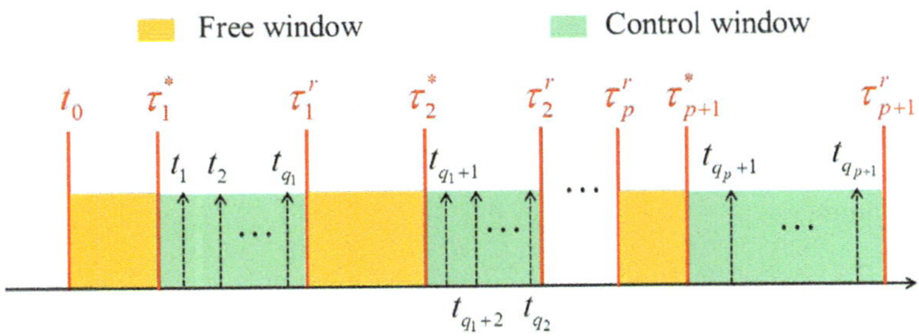

Fig. 2.15 Diagram of intermittent impulsive control

Theorem 2.5 Suppose the Assumptions 1.1 and 1.2 hold, if

$$(\lambda_A + 2l)(\tau_p^r - \tau_{p-1}^r) + \ln(\bar{\lambda}_p \xi) \leq 0, \tag{2.69}$$

where $\xi > 1$, λ_A and $\bar{\lambda}_p$ have the same definitions in Theorem 2.4, then the consensus of MASs will be achieved.

Proof Choose the Lyapunov function $V(t) = \delta^T \delta$, similar to the proof of Theorem 2.4, one can obtain when $t = \tau_1^*$,

$$V(\tau_1^*) \leq V(t_0) \exp((\lambda_A + 2l)(\tau_1^* - t_0)). \tag{2.70}$$

When $t = \tau_1^r$,

$$\begin{aligned} V(\tau_1^r) &\leq V(\tau_1^*)\lambda_1\lambda_2 \cdots \lambda_{q_1} \exp((\lambda_A + 2l)(\tau_1^r - \tau_1^*)) \\ &\leq V(t_0)\lambda_1\lambda_2 \cdots \lambda_{q_1} \exp((\lambda_A + 2l)(\tau_1^r - t_0)). \end{aligned} \tag{2.71}$$

From (2.56), one gets

$$V(\tau_2^*) \leq V(\tau_1^r) exp((\lambda_A + 2l)(\tau_1^* - \tau_1^r)). \tag{2.72}$$

When $t = \tau_2^r$,

$$\begin{aligned} V(\tau_2^r) &\leq V(\tau_2^*)\lambda_{q_1+1}\lambda_{q_1+2} \cdots \lambda_{q_2} \exp((\lambda_A + 2l)(\tau_2^r - \tau_2^*)) \\ &\leq V(\tau_1^r)\lambda_{q_1+1}\lambda_{q_1+2} \cdots \lambda_{q_2} \exp((\lambda_A + 2l)(\tau_2^r - \tau_1^r)) \\ &\leq V(t_0)\lambda_1\lambda_2 \cdots \lambda_{q_1} \exp((\lambda_A + 2l)(\tau_1^r - t_0)) \\ &\quad \lambda_{q_1+1}\lambda_{q_1+2} \cdots \lambda_{q_2} \exp((\lambda_A + 2l)(\tau_2^r - \tau_1^r)). \end{aligned} \tag{2.73}$$

In general, when $t = \tau_p^r$,

$$
\begin{aligned}
&V(\tau_p^r)\\
&\le V(\tau_p^*)\lambda_{q_{p-1}+1}\lambda_{q_{p-1}+2}\cdots\lambda_{q_p}\exp((\lambda_A+2l)(\tau_p^r-\tau_p^*))\\
&\le V(t_0)\lambda_1\lambda_2\cdots\lambda_{q_1}\exp((\lambda_A+2l)(\tau_1^r-t_0))\lambda_{q_1+1}\lambda_{q_1+2}\\
&\quad\cdots\lambda_{q_2}\exp((\lambda_A+2l)(\tau_2^r-\tau_1^r))\cdots\\
&\quad\lambda_{q_{p-1}+1}\lambda_{q_{p-1}+2}\cdots\lambda_{q_p}\exp((\lambda_A+2l)(\tau_{p+1}^r-\tau_p^r))\\
&=V(t_0)\bar{\lambda}_1\exp((\lambda_A+2l)(\tau_1^r-t_0))\bar{\lambda}_2\exp((\lambda_A+2l)(\tau_2^r-\tau_1^r))\\
&\quad\cdots\bar{\lambda}_p\exp((\lambda_A+2l)(\tau_{p+1}^r-\tau_p^r)).
\end{aligned}
\tag{2.74}
$$

From (2.69) in Theorem 2.5, one gets

$$
V(t)\le V(t_0)\left(\frac{1}{\xi}\right)^{p-1}. \tag{2.75}
$$

Since $V(t_0)$ is bounded constant, and $\lim\limits_{p\to\infty} 1/\xi^p=0$, then $\lim\limits_{p\to\infty} V(\tau_p^r)=0$.

When $t\in(\tau_p^r,\tau_{p+1}^*]$, from (2.56),

$$
\begin{aligned}
V(t)&\le V(\tau_{p+1}^*)\\
&\le V(\tau_p^r)\exp((\lambda_A+2l)(\tau_{p+1}^*-\tau_p^r)).
\end{aligned}
\tag{2.76}
$$

When $t\in(\tau_{p+1}^*,t_{q_p+1}]$, from (2.56) and (2.76),

$$
\begin{aligned}
V(t)&\le V(t_{q_p+1})\\
&\le V(\tau_{p+1}^*)\exp((\lambda_A+2l)(t_{q_p+1}-\tau_{p+1}^*))\\
&\le V(\tau_p^r)\exp((\lambda_A+2l)(t_{q_p+1}-\tau_p^r)).
\end{aligned}
\tag{2.77}
$$

When $t\in(t_{q_p+1},t_{q_p+2}]$, from (2.56), (2.57), and (2.77),

$$
\begin{aligned}
V(t)&\le V(t_{q_p+2})\\
&\le \lambda_{q_p+1}V(t_{q_p+1})\exp((\lambda_A+2l)(t_{q_p+2}-t_{q_p+1}))\\
&\le \lambda_{q_p+1}V(T_p)\exp((\lambda_A+2l)(t_{q_p+2}-\tau_p^r)).
\end{aligned}
\tag{2.78}
$$

In general, when $t\in(t_{q_{p+1}},\tau_{p+1}^r]$

$$
\begin{aligned}
V(t)&\le V(\tau_{p+1}^r)\\
&\le \lambda_{q_p+1}\lambda_{q_p+2}\cdots\lambda_{q_{p+1}}V(T_p)\exp((\lambda_A+2l)(\tau_{p+1}^r-\tau_p^r)).
\end{aligned}
\tag{2.79}
$$

Since λ_{q_p+1}, $\lambda_{q_p+2},\ldots,$ $\lambda_{q_{p+1}}$, $\exp((\lambda_A+2l)(\tau_{p+1}^*-\tau_p^r))$, $\exp((\lambda_A+2l)(t_{q_p+1}-\tau_p^r))$, $\exp((\lambda_A+2l)(t_{q_p+2}-\tau_p^r)),\ldots,$ $\exp((\lambda_A+2l)(\tau_{p+1}^r-\tau_p^r))$ are bounded constants, and $\lim\limits_{p\to\infty} V(\tau_p^r)=0$, then one gets $\lim\limits_{t\to\infty} V(t)=0$. Therefore, the consensus of MASs is realized under the intermittent impulsive protocol. This completes the proof.

Remark 2.11 Similarly to the variable impulsive protocol, the intermittent impulsive control strategy in Theorem 2.5 does not need to fix any impulsive instant. The following steps are given to explain how to use Theorem 2.5.
Step 1: Determine the begin and end of each control window, that is, determine the instants of τ_p^* and $\tau_p^r (p \in \mathbb{N}^+)$.
Step 2: From (2.69) in Theorem 2.5, calculate the parameter $\bar{\lambda}_p$ as

$$\bar{\lambda}_p \leq \exp(-(\lambda_A + 2l)(\tau_p^r - \tau_{p-1}^r))/\xi.$$

Step 3: Determine b_k and the number of impulses which need to be exerted in each control window.
Step 4: In each control window, exert the required impulses at arbitrary impulsive instants.

Remark 2.12 If the control gain of each impulsive control is chosen as a constant b, then $\bar{\lambda}_p = \lambda^{q_p - q_{p-1}}$, where λ is the maximum eigenvalue of $(b(L + C) + I_N)^T (b(L + C) + I_N)$. From (2.69) in Theorem 2.5, the number of impulses in control window (τ_p^*, τ_p^r) is

$$\begin{aligned} q_p - q_{p-1} &= \frac{\ln \bar{\lambda}_p}{\ln \lambda} \\ &\geq \frac{\ln \frac{\exp(-(\lambda_A + 2l)(\tau_p^r - \tau_{p-1}^r))}{\xi}}{\ln \lambda}. \end{aligned} \tag{2.80}$$

Then the number of impulsive control in pth control window can be chosen as $\lceil \frac{\ln \exp(-(\lambda_A + 2l)(\tau_p^r - \tau_{p-1}^r)) - \ln \xi}{\ln \lambda} \rceil$, where $\lceil \cdot \rceil$ is the ceil operator. Therefore, if the length of $\tau_p^r - \tau_{p-1}^r$ is invariant, then the number of impulses in (τ_p^*, τ_p^r) is invariant whether τ_p^* changes or not. In addition, from (2.80), one knows λ affects the number of impulses in the control window (τ_p^*, τ_p^r). If λ decreases, $q_p - q_{p-1}$ decreases. Therefore, to exert less impulses in control window, it is necessary to select the appropriate impulsive control gain b to decrease the value of λ.

Remark 2.13 Note that the value of τ_p^* determines the ratio of the free and control windows in a period. When the length of $\tau_p^r - \tau_{p-1}^r$ is invariant and the ratio of free window (τ_{p-1}^r, τ_p^*) becomes larger, then the length of control window (τ_p^*, τ_p^r) becomes smaller. Since the number of impulses in (τ_p^*, τ_p^r) is invariant, the density of impulsive control becomes larger when the ratio of free window increases. In short, the selection of parameter τ_p^* and τ_p^r is a trade-off between obtaining a longer relax interval and obtaining a smaller impulsive density.

2.4 Simulation Results

In this section, some examples are provided to validate the effectiveness of the variable and intermittent impulsive methods in this chapter.

Chua's oscillator (Shil'nikov 1993) is described as

$$\begin{cases} \dot{x}_{i1} = p_1(x_{i2} - x_{i1} - g(x_{i1})), \\ \dot{x}_{i2} = x_{i1} - x_{i2} + x_{i3}, \\ \dot{x}_{i3} = -p_2 x_{i2}, \end{cases}$$

where p_1 and p_2 are system parameters and the $g(\cdot)$ is the piecewise linear function, which is described as

$$g(x_{i1}) = m_2 x_{i1} + 0.5(m_1 - m_2)(|x_{i1} + 1| - |x_{i1} - 1|),$$

where $m_1 < m_2 < 0$ are two given constants, then

$$A = \begin{bmatrix} -p_1(1 + m_2) & p_1 & 0 \\ 1 & -1 & 1 \\ 0 & -p_2 & 0 \end{bmatrix},$$

$$f(x_i) = \begin{bmatrix} -0.5p_1(m_1 - m_2)(|x_{i1} + 1| - |x_{i1} - 1|) \\ 0 \\ 0 \end{bmatrix}.$$

Let $p_1 = 9.21$, $p_2 = 15.995$, $m_1 = -1.25$, $m_2 = -0.758$. Then it yields $\lambda_A = 16.5492$, $l = |(m_1 - m_2)p_1| = 4.5313$. Consider the topology of MASs as shown in Fig. 2.16. The main results (Theorems 2.1–2.5) will be validated based on the topology.

From Fig. 2.16, the adjacency matrix, in-degree matrix, Laplacian matrix, and pinning gain matrix are obtained as follows:

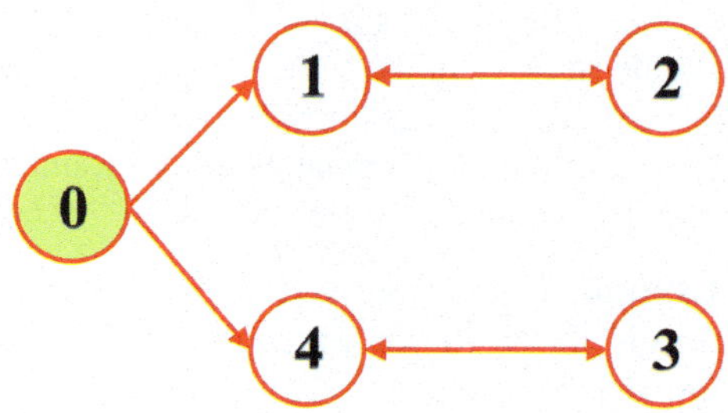

Fig. 2.16 Topology graph of the multi-agent systems

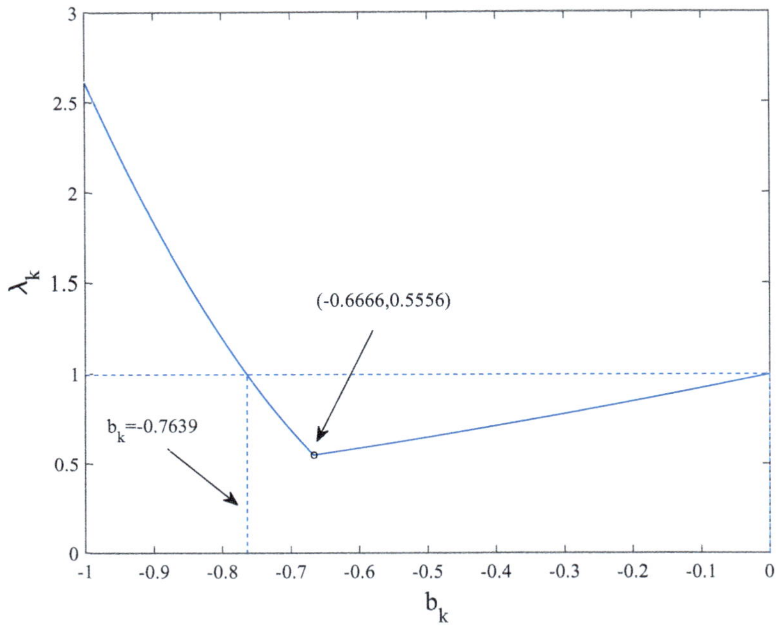

Fig. 2.17 Allowable range of impulsive control gain b_k

$$\mathcal{A} = \begin{bmatrix} 0&1&0&0 \\ 1&0&0&0 \\ 0&0&0&1 \\ 0&0&1&0 \end{bmatrix}, \quad D = \begin{bmatrix} 1&0&0&0 \\ 0&1&0&0 \\ 0&0&1&0 \\ 0&0&0&1 \end{bmatrix},$$

$$L = D - \mathcal{A} = \begin{bmatrix} 1&-1&0&0 \\ -1&1&0&0 \\ 0&0&1&-1 \\ 0&0&-1&1 \end{bmatrix}, \quad C = \begin{bmatrix} 1&0&0&0 \\ 0&0&0&0 \\ 0&0&0&0 \\ 0&0&0&1 \end{bmatrix}.$$

From the discussion in Remark 2.3, the choice of impulsive control gain b_k should make sure $\lambda_k \in (0, 1)$. The selectable range of b_k is presented in Fig. 2.17. From Fig. 2.17, one knows the allowable range of b_k is $(-0.7639, 0)$, and when $b_k = -0.6666$, λ_k is smallest. From Remark 2.3, to have a larger selectable range of impulsive intervals, we choose $b_k = -0.6666$ in this chapter.

Example 2.1 (*Impulsive time window*) Let $b_k = -0.666$, $\xi = 1.01$, from the consensus Eq. (2.10) in Theorem 2.1, we can obtain the left endpoint distance of impulsive window as

$$\tau_{k+1}^l - \tau_k^l = \Delta_{k+1} - \mu_{k+1} \leq -\frac{\ln(\lambda_k \xi)}{\lambda_A + 2l} = -\frac{\ln(1.01\lambda_k)}{25.6118}.$$

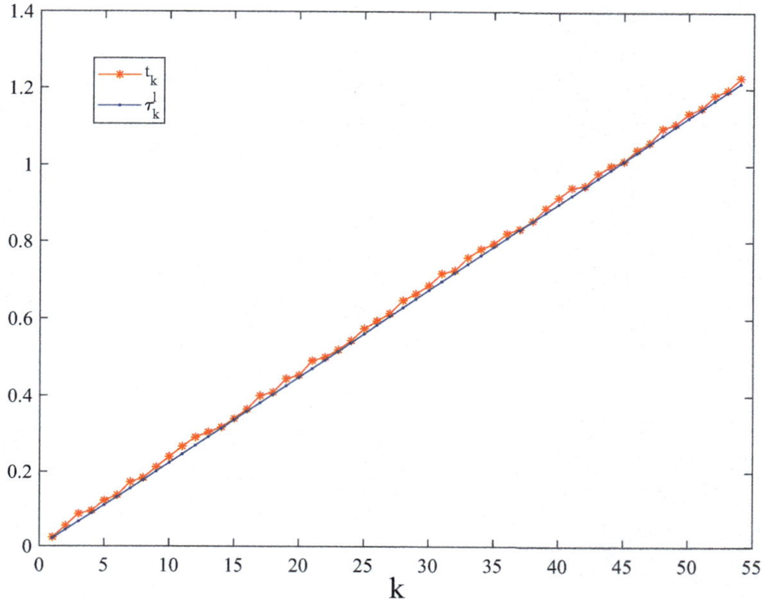

Fig. 2.18 The relation between t_k and τ_k^l

Select the initial values of each agent in MASs as $x_0 = [-2, 2, 3]^T$, $x_1 = [0, 1, 1]^T, x_2 = [-1, 1, 1]^T, x_3 = [-1, 1, 1]^T, x_4 = [0, 1, 1]^T$. The relation between the impulsive instant t_k and left endpoint τ_k^l of kth impulsive window is shown in Fig. 2.18. It is apparent that the actual impulsive instant t_k is greater than the left endpoint instant τ_k^l. Furthermore, Fig. 2.19 presents the curve of obtained impulsive interval $t_k - t_{k-1}$ by using Theorem 2.1. From Fig. 2.19, one gets the impulsive is variable by using the method of impulsive time window. Figure 2.20 shows that all agents can realize the consensus of the leader.

Now keep b_k and ξ unchanged, from (2.22), one gets the center distance of two adjacent impulsive time window as

$$\tau_{k+1} - \tau_k \leq -\frac{\ln(\lambda_k \xi)}{\lambda_A + 2l} = -\frac{\ln(1.01\lambda_k)}{25.6118}.$$

The relation between the impulsive instant t_k and the center of impulsive time window τ_k is shown in Fig. 2.21, which shows red star nodes (impulsive instants t_k) are distributed at both sides of blue points (the centers of impulsive time window). The obtained impulsive interval via Theorem 2.2 is shown in Fig. 2.22. Figure 2.23 shows that consensus of the MASs can be realized.

Example 2.2 (*Odd impulsive control*) Next we consider the case of the odd impulsive sequence. Keep b_k and ξ unchanged, the initial conditions are taken as

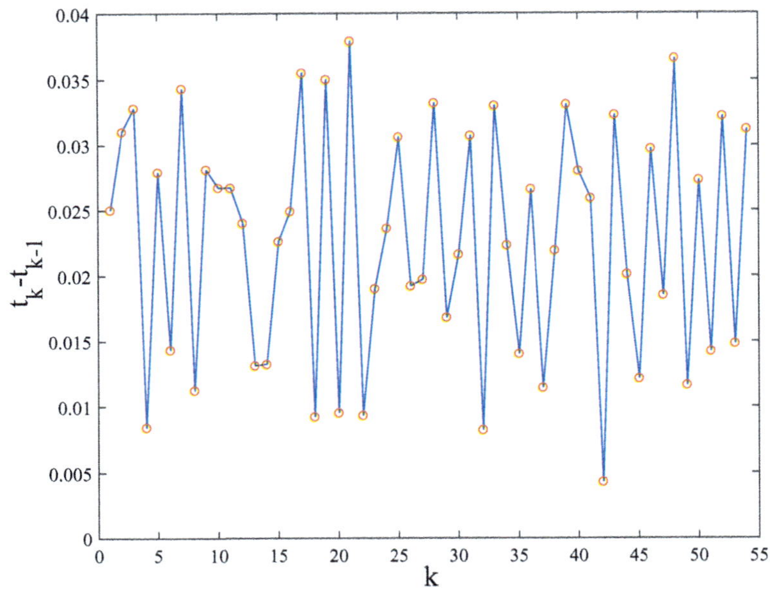

Fig. 2.19 Impulsive interval $t_k - t_{k-1}$ versus k via Theorem 2.1

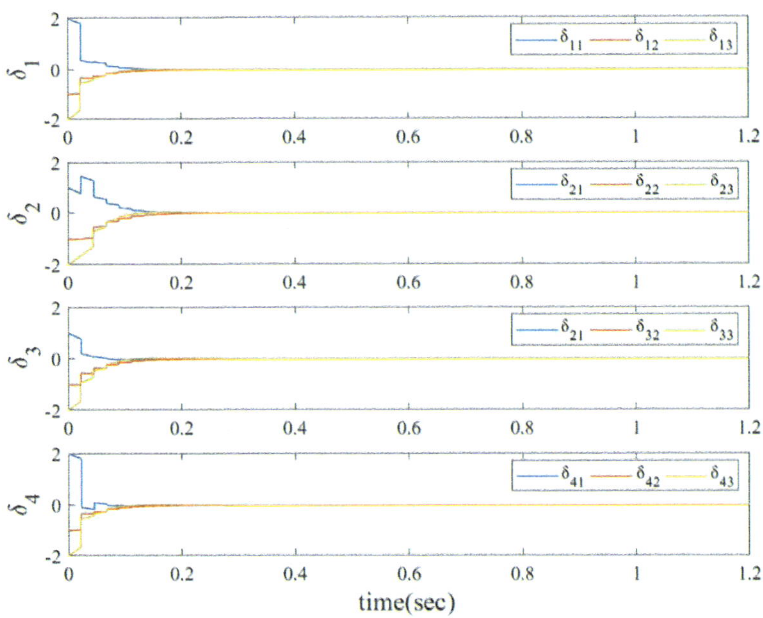

Fig. 2.20 The global consensus errors in Theorem 2.1

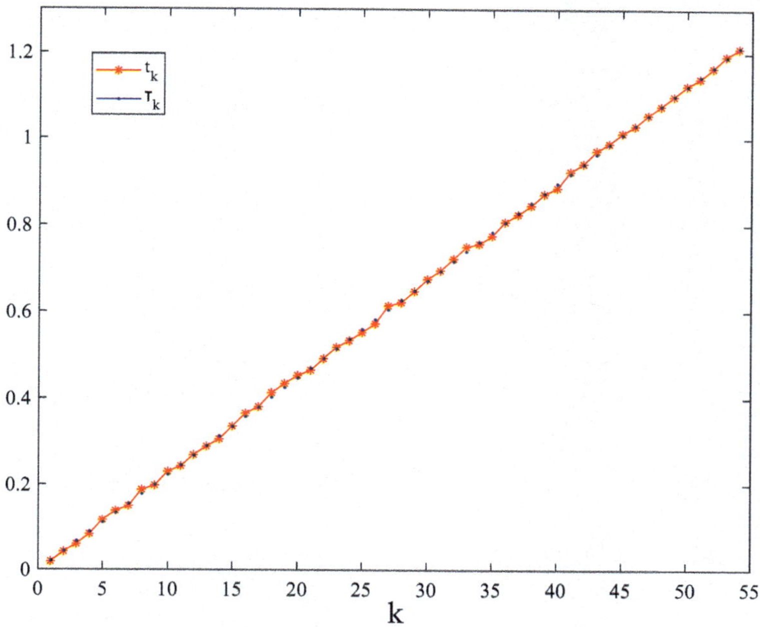

Fig. 2.21 The relation between t_k and τ_k

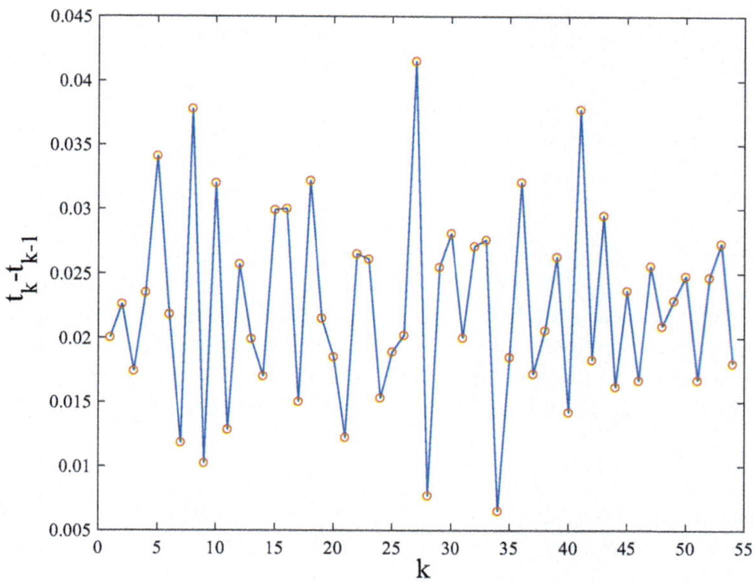

Fig. 2.22 Impulsive interval $t_k - t_{k-1}$ versus k via Theorem 2.2

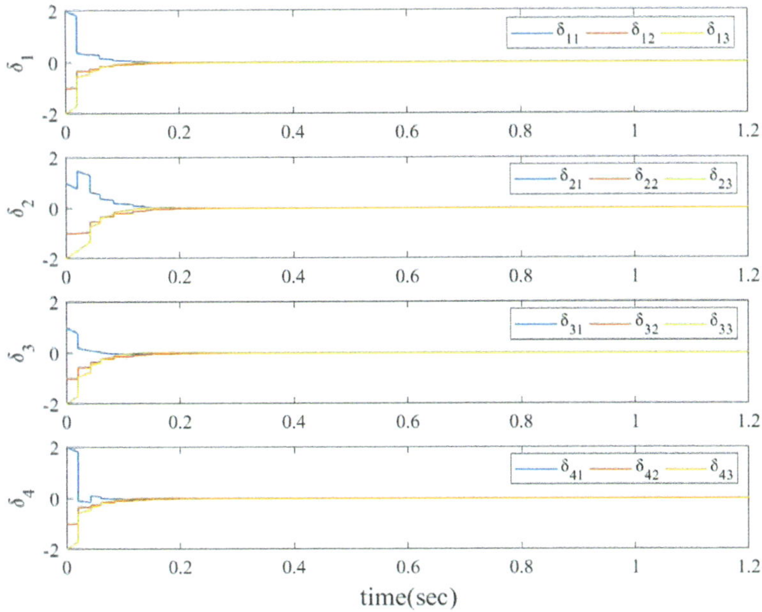

Fig. 2.23 The global consensus errors in Theorem 2.2

$x_0 = [2, 0, -1]^T$, $x_1 = [-3, 1, 2]^T$, $x_2 = [-2, -1, 2]^T$, $x_3 = [-1, -1, 1]^T$, $x_4 = [1, 1, 1]^T$.

From (2.41) in Theorem 2.3, one can acquire

$$t_{2k+1} - t_{2k-1} \leq -\frac{\ln(\lambda_{2k-1}\lambda_{2k}\xi)}{\lambda_A + 2l} = -\frac{\ln(1.01\lambda_{2k-1}\lambda_{2k})}{25.6118}.$$

Thus, we can get the odd impulsive sequence $\{t_{2k-1}\}$, which is presented with blue mark in Fig. 2.24. The curve of impulsive interval in this example is presented in Fig. 2.25, which also shows the impulsive interval is variable. Based on the characteristics of odd impulsive protocol, the corresponding even impulsive instants t_{2k} can be chosen within the interval (t_{2k-1}, t_{2k+1}), which is also presented with red mark in Fig. 2.24. As shown in Fig. 2.26, all the agents can synchronize to the leader in 0.16s.

Example 2.3 (*A more flexible variable impulsive control protocol*) Select the initial values of each agent in MASs as $x_0 = [-2, 2, 1]^T$, $x_1 = [-3, 3, 5]^T$, $x_2 = [-2, 1, 0]^T$, $x_3 = [-1, -1, -2]^T$, $x_4 = [1, 1, -3]^T$. Choose the right endpoint of each window as $T_1 = 0.3$, $T_2 = 0.65$, $T_3 = 0.9$, $T_4 = 1.25$. Choose $\xi = 1.01$, then the global consensus error of four follower agents is presented in Fig. 2.27. Then one gets the consensus of MASs can be realized.

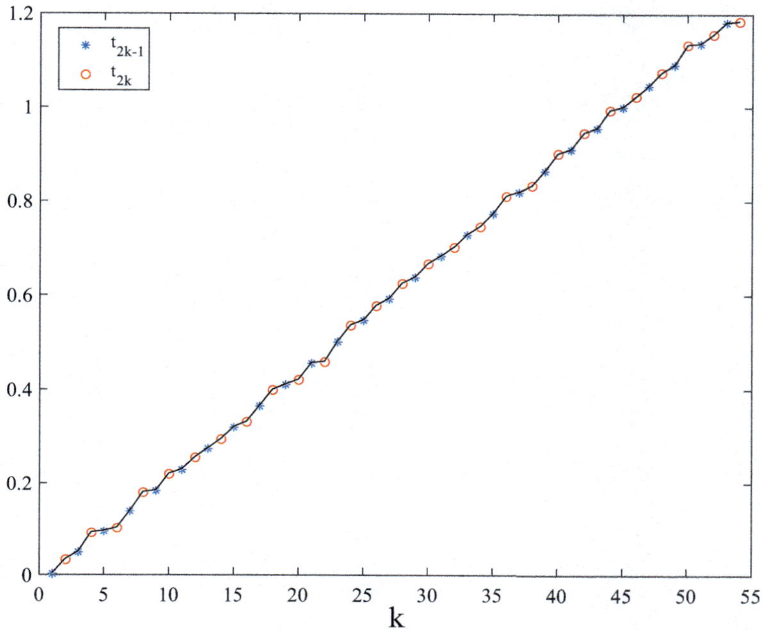

Fig. 2.24 The relation between t_k and k

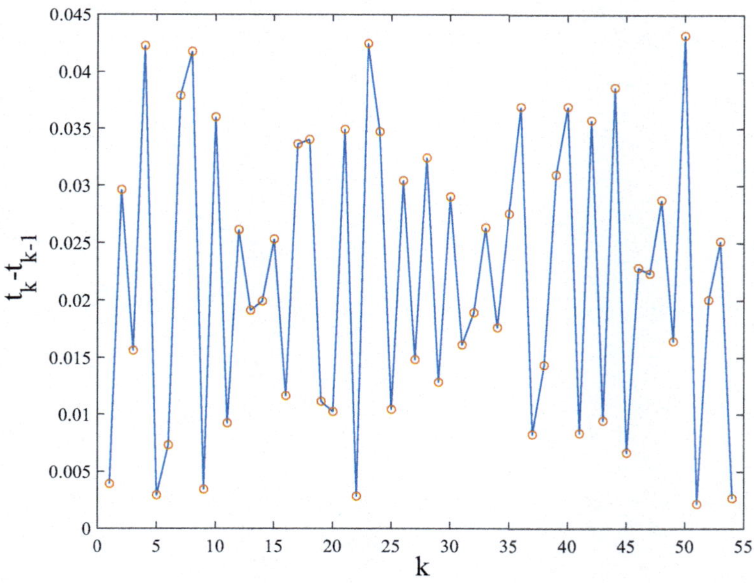

Fig. 2.25 Impulsive interval $t_k - t_{k-1}$ versus k via Theorem 2.3

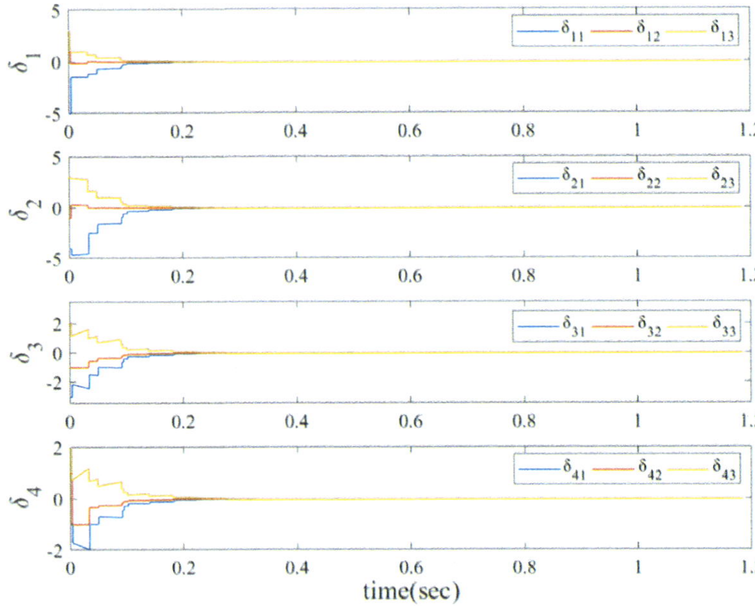

Fig. 2.26 The global consensus errors in Theorem 2.3

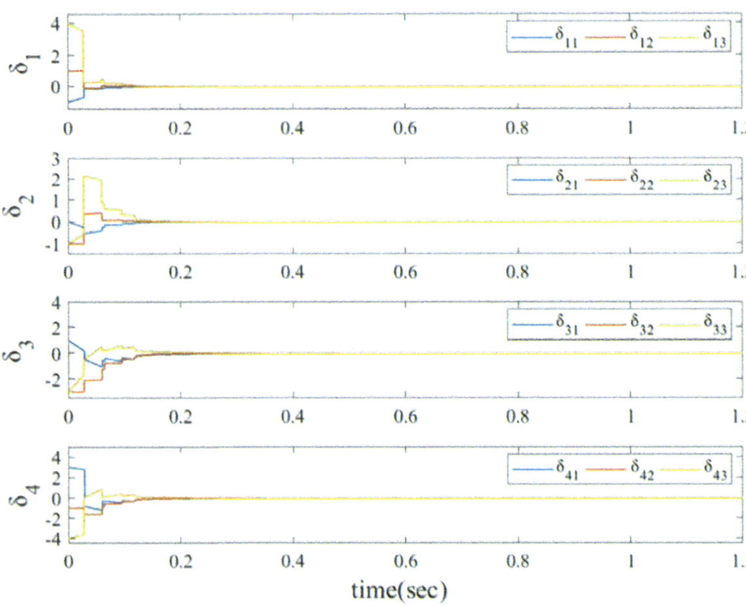

Fig. 2.27 The global consensus errors in Theorem 2.4

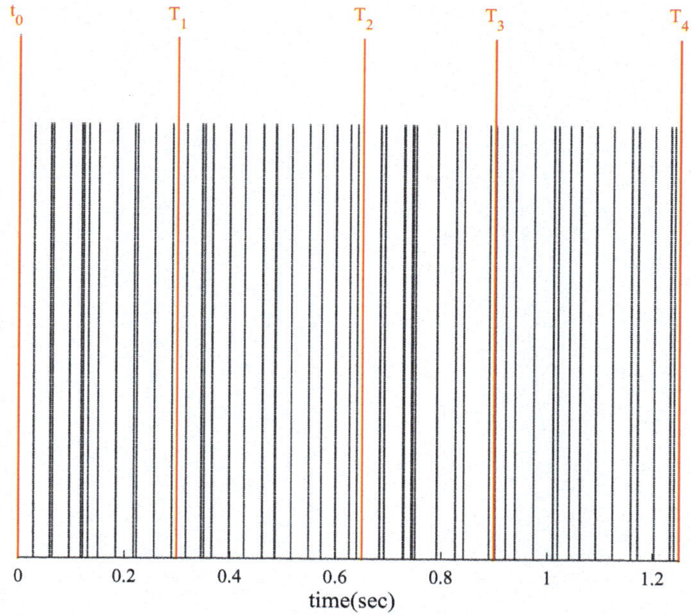

Fig. 2.28 Diagram of variable impulsive control in Theorem 2.4

Figure 2.28 shows the diagram of variable impulsive control in Example 2.3. In Fig. 2.28, the solid red lines represent the right endpoint of each window, and the dotted black lines indicate impulsive instants. The impulsive interval of variable impulsive control is presented in Fig. 2.29. It is obvious that variable impulsive control is flexible and does not require fixed impulsive interval and impulsive instant.

It should be noted that the values of system model parameters, topology structure, and impulsive control gain b_k in Example 2.3 are the same as those in Examples 2.1 and 2.2. Comparing Fig. 2.29 with Figs. 2.19 and 2.22, one knows that the variable impulsive scheme designed by Theorem 2.4 gives a wider range of impulsive intervals. For example, in Fig. 2.29, the upper bound of the impulsive interval can be close to $0.05s$. This is because the variable impulsive control scheme designed by Theorem 2.4 does not need to limit the impulsive instants to a small time window, so the application of impulses is more flexible, and the obtained range of impulsive interval will be larger. Comparing Figs. 2.29 to 2.25, we know that the variable impulsive control scheme designed by Theorem 2.4 allows a wider range of the impulsive interval than odd impulsive control. This is because the variable impulsive control scheme designed by Theorem 2.4 does not need to determine any odd or even impulsive instant, which is more flexible than odd impulsive control protocol.

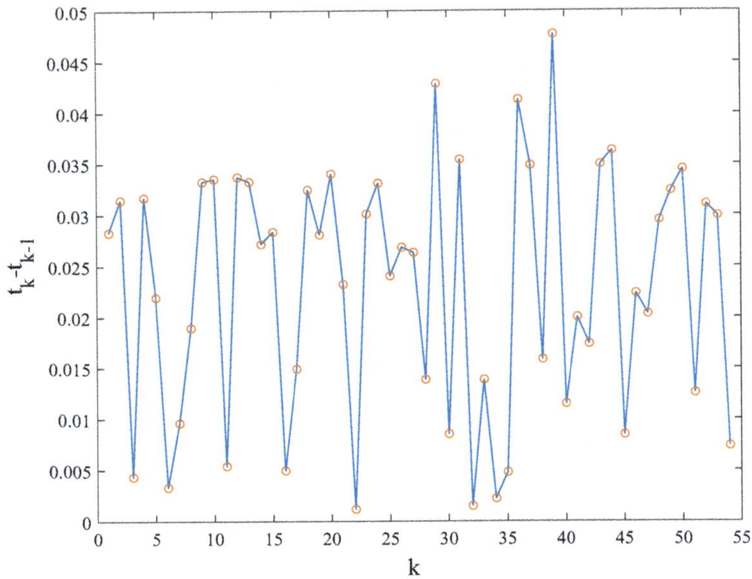

Fig. 2.29 Impulsive interval τ_k versus k in Theorem 2.4

Example 2.4 (*Intermittent impulsive control protocol*) Consider the same system model and topology as in Example 2.3, in this subsection, an example verifies the validity of the intermittent impulsive protocol. In this example, we choose the initial values of each agent as $x_0 = [-2, 3, 4]^T$, $x_1 = [-3, 2, 0]^T$, $x_2 = [1, -1, 5]^T$, $x_3 = [2, -1, 2]^T$, $x_4 = [1, 0, 3]^T$.

Choose $\tau_1^* = 0.15s$, $\tau_1^r = 0.3s$, $\tau_2^* = 0.5s$, $\tau_2^r = 0.65s$, $\tau_3^* = 0.75s$, $\tau_3^r = 0.9s$, $\tau_4^* = 1s$, $\tau_4^r = 1.25s$. Keeping the value of b_k and ξ unchanged, the states of different nodes are presented in Fig. 2.30.

It is not difficult to find that $\tau_2^r - \tau_1^r = \tau_4^r - \tau_3^r = 0.35s$. In the second period, the ratio of the control window is $0.15/0.35 = 42.9\%$. In the second period, the ratio of the control window is $0.25/0.35 = 71.4\%$. Because the ratio of the control window in the second period is smaller than the fourth period, the density of impulsive control in second period is larger than the fourth period, which is presented in Fig. 2.31.

Figure 2.32 presents the impulsive interval in this example. There are some points with larger impulsive intervals, which indicates that a free window has just been experienced. In addition, Fig. 2.32 shows that the intermittent impulsive protocol allows for larger impulsive intervals, which cannot be realized in both impulsive time window and odd impulsive control strategy.

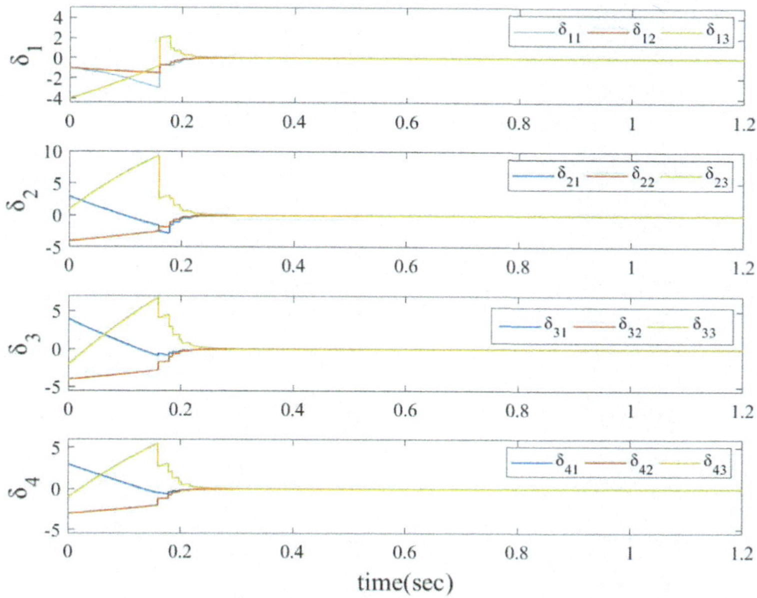

Fig. 2.30 The global consensus errors in Theorem 2.5

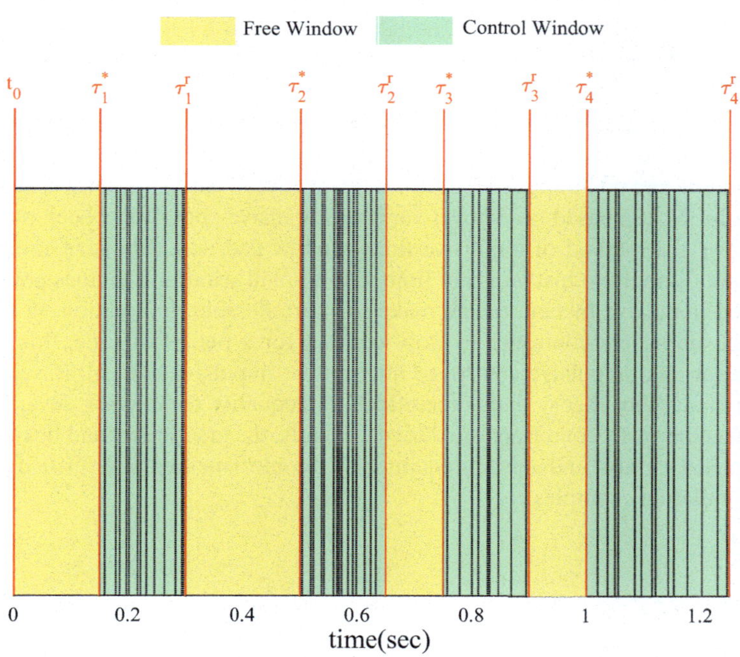

Fig. 2.31 Diagram of intermittent impulsive control in Theorem 2.5

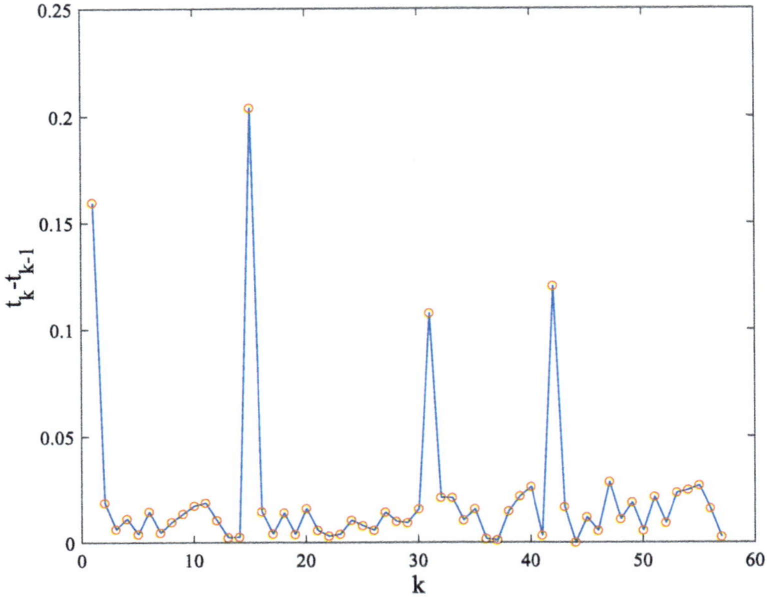

Fig. 2.32 Impulsive interval τ_k versus k in Theorem 2.5

2.5 Summary

In this chapter, the consensus scheme of nonlinear MASs via variable impulsive control is studied. Firstly, we consider the distributed control protocol with impulsive time window and odd impulsive sequence, respectively. Next, the consensus control scheme for MASs based on a novel variable impulsive control has been proposed. Comparing the method of impulsive time window and odd impulsive control, the selection of impulsive instant is not limited to a small window and does not require any fixed impulsive instant, which makes it more flexible. In addition, in order to allow the impulsive controller to stop working for a period of time, this chapter extends variable impulsive control to intermittent impulsive control. Based on the Lyapunov stability theory and some effective inequality techniques, several novel sufficient consensus conditions are derived. Finally, the practicality and flexibility of the proposed variable and intermittent impulsive control methods are verified through some simulation examples.

References

Chen F, Wang H, Li C (2015) Impulsive control of memristive chaotic systems with impulsive time window. Math Probl Eng 2015:927327

Feng Y, Li C, Huang T (2015) Uniformly stability of delay-dependent linear systems with impulse time windows. In: Chinese Control Conference (CCC), pp 618–622

Li H, Li C, Huang T (2017) Periodicity and stability for variable-time impulsive neural networks. Neural Netw 94:24–33

Long T, Yang S, Wang Q, Liao X (2021) Finite-time consensus of nonlinear multi-agent systems via impulsive time window theory: a two-stage control strategy. Nonlinear Dyn 105(4):3285–3297

Shil'nikov LP (1993) Chua's circuit: rigorous results and future problems. IEEE Trans Circuits Syst I: Fundam Theory Appl 40(10):784–786

Wang X, Yu J, Li C, Wang H, Huang T, Huang J (2015) Robust stability of stochastic fuzzy delayed neural networks with impulsive time window. Neural Netw 67:84–91

Wang H, Duan S, Li C, Wang L, Huang T (2016a) Globally exponential stability of delayed impulsive functional differential systems with impulse time windows. Nonlinear Dyn 84(3):1655–1665

Wang H, Duan S, Li C, Wang L, Huang T (2016b) Stability criterion of linear delayed impulsive differential systems with impulse time windows. Int J Control Autom Syst 14(1):174–180

Wang X, Wang H, Li C, Huang T (2016c) Synchronization of coupled delayed switched neural networks with impulsive time window. Nonlinear Dyn 84(3):1747–1757

Wang K, Tang Z, Park JH, Feng J (2022) Impulsive time window based quasi-consensus on stochastic nonlinear multi-agent systems. IEEE Trans Netw Sci Eng 9(5):3602–3613

Wei L, Liao Y, Luo S, Chen W, Huang G (2023) Consensus of nonlinear multi-agent systems via distributed hybrid controls. Int J Robust Nonlinear Control 33(13):7510–7531

Wu H, Xiong J, Feng Y, Tu Z (2020) Lag synchronization of memristor oscillator systems via a double impulsive controller with time windows. Zidonghua Xuebao/Acta Automatica Sinica 46(7):1507–1516

Xu X, Zhang Z, Li C, Zhang Q (2020) A hybrid protocol for the average consensus of multi-agent systems with impulse time window. J Frankl Inst 357(7):4222–4241

Zhou Y, Li C, Huang T, Wang X (2017) Impulsive stabilization and synchronization of Hopfield-type neural networks with impulse time window. Neural Comput Appl 28(4):775–782

Zhou Y, Li C, Chen L, Huang T (2018) Global exponential stability of memristive Cohen-Grossberg neural networks with mixed delays and impulse time window. Neurocomputing 275:2384–2391

Zou L, Peng Y, Feng Yu, Tu Z (2018) Impulsive control of nonlinear systems with impulse time window and bounded gain error. Nonlinear Anal: Modell Control 23(1):40–49

Chapter 3
Distributed Asynchronous Impulsive Consensus of Multi-agent Systems

3.1 Introduction

Due to the hardware constraints, the synchronized impulsive protocol is difficult to be applied in large-scale networks. To overcome the limitation of synchronized impulsive method for MASs, some researchers have done research on asynchronous impulsive control. Under the asynchronous impulsive control protocol, only a part of nodes are controlled at one impulsive instant, so only a part of nodes need to interact with their neighboring nodes, resulting in lower information transmission and control consumption at each impulsive instant. Jiang et al. (2018) discussed the asynchronous impulsive consensus of the leaderless second-order MASs. Han et al. (2019) studied the leaderless consensus of discrete MASs by the event-based asynchronously impulsive control. Zhu et al. (2019) investigated the asynchronous impulsive control for the consensus of the leaderless linear MASs with switching topologies. Liu et al. (2019) adopted the asynchronous impulsive control to achieve the leaderless consensus of second-order MASs. Zhang et al. (2022) discussed how MASs with a leader can achieve distributed bipartite tracking consensus using asynchronous impulsive control strategies. Gong et al. (2021) investigated the bounded leader-following consensus of heterogeneous directed delayed MASs via asynchronous impulsive control. Zhou et al. (2022) researched asynchronous consensus of heterogeneous second-order MASs on time scales. Li et al. (2024) realized the interval consensus of second-order MASs via asynchronous impulsive protocol.

In this chapter, by designing the asynchronous impulsive protocol, some novel sufficient consensus conditions are obtained. Firstly, the design of asynchronous impulsive controller is completed, that is, the consensus control objective can be achieved by controlling a part of agents. Then, based on impulsive differential equations and the Lyapunov stability theory, a sufficient condition for asynchronous impulsive consensus control is obtained. Finally, this chapter proposes two schemes to reduce the conservatism of the asynchronous impulsive control method, thereby enhancing the practicability of the control method.

Z. Zhang et al., *Distributed Impulsive Coordination of Multi-Agent Systems*, Intelligent Control and Learning Systems 15, https://doi.org/10.1007/978-981-97-6262-0_3

3.2 Problem Description

Consider the follower nodes in MASs as

$$\dot{x}_i(t) = Ax_i(t) + f(x_i(t)) + u_i(t), \ i = 1, 2, \ldots N, \tag{3.1}$$

where $x_i = [x_{i1}, x_{i2}, \ldots, x_{in}]^T \in \mathbb{R}^n$ is the state vector of agent i, $f : \mathbb{R}^n \to \mathbb{R}^n$ is the nonlinear function of node i. $A \in \mathbb{R}^{n\times n}$ is a known system matrix.

Consider the leader agent as

$$\dot{x}_0(t) = Ax_0(t) + f(x_0(t)), \tag{3.2}$$

where $x_0 = [x_{01}, x_{02}, \ldots, x_{0n}]^T \in \mathbb{R}^n$ is the state vector of leader node.

From agent 1 to agent N in $\mathcal{G}$, one can construct the set of controlled agents when $t = t_k$ as $Z(k) = \{z_1(k), z_2(k), \ldots, z_{p(k)}(k)\}$, where $p(k)$ is the number of controlled agents at impulsive instant t_k. For example, in Fig. 3.1, $N = 4$, the blue nodes 1, 3, and 4 are the agents controlled at impulsive instant t_k, then $z_1(k) = 1$, $z_2(k) = 3$, $z_3(k) = 4$, and $p(k) = 3$. The controlled agents set $Z(k)$ is very important in the subsequent analysis and proof process.

Definition 3.1 The global synchronization error of the agent i in MASs is defined as

$$\delta_i(t) = x_i(t) - x_0(t), \tag{3.3}$$

where $\delta_i(t) = [\delta_{i1}(t), \delta_{i2}(t), \ldots, \delta_{in}(t)]^T \in \mathbb{R}^n$.

Definition 3.2 The synchronization error of local neighborhood agents for the node i in MASs is defined as

$$e_i(t) = \sum_{j\in N_i} a_{ij}(x_i(t) - x_j(t)) + c_i(x_i(t) - x_0(t)), \tag{3.4}$$

where $e_i(t) = [e_{i1}(t), e_{i2}(t), \ldots, e_{in}(t)]^T \in \mathbb{R}^n$.

Define the subgraph of $\mathcal{G}$, which only contains the controlled agents at impulsive instants t_k as $\tilde{\mathcal{G}}_k = (\tilde{\mathcal{V}}_k, \tilde{\mathcal{E}}_k, \tilde{\mathcal{A}}_k)$, where $\tilde{\mathcal{V}}_k = \{v_{z_1(k)}, v_{z_2(k)}, \ldots, v_{z_{p(k)}(k)}\}$, $\tilde{\mathcal{E}}_k \in (\tilde{\mathcal{V}}_k \times \tilde{\mathcal{V}}_k)$, and $\tilde{\mathcal{A}}_k \in \mathbb{R}^{p(k)\times p(k)}$.

Define the augmented graph as $\bar{\mathcal{G}} = (\bar{\mathcal{V}}, \bar{\mathcal{E}}, \bar{\mathcal{A}})$, where $\bar{\mathcal{V}} = \{v_0, v_1, \ldots, v_N\}$, $\mathcal{E} \in (\bar{\mathcal{V}} \times \bar{\mathcal{V}})$, and $\bar{\mathcal{A}} = [\bar{a}_{ij}] \in \mathbb{R}^{(N+1)\times(N+1)}$. Then we define the augmented graph which only contains a leader agent and the controlled follower agents at impulsive instants t_k as $\bar{\tilde{\mathcal{G}}}_k = (\bar{\tilde{\mathcal{V}}}_k, \bar{\tilde{\mathcal{E}}}_k, \bar{\tilde{\mathcal{A}}}_k)$, where $\bar{\tilde{\mathcal{V}}}_k = \{v_0, v_{z_1(k)}, \ldots, v_{z_{p(k)}(k)}\}$, $\bar{\tilde{\mathcal{E}}}_k \in (\bar{\tilde{\mathcal{V}}}_k \times \bar{\tilde{\mathcal{V}}}_k)$ and $\bar{\tilde{\mathcal{A}}}_k \in \mathbb{R}^{(p(k)+1)\times(p(k)+1)}$.

For the leader-following consensus problem, it is necessary to give the following assumption about the topological structure.

Assumption 3.1 The graphs $\bar{\mathcal{G}}$ and $\bar{\tilde{\mathcal{G}}}_k$ for all impulsive instants should contain a spanning tree for which the root is the leader agent.

Remark 3.1 Assumption 3.1 satisfies almost any possible topological structure, such as the graphs $\bar{\mathcal{G}}$ and $\bar{\bar{\mathcal{G}}}_k$ which contain a spanning tree, and at least the root agent can receive information of the leader agent. It is worth to note that the directed graphs $\bar{\mathcal{G}}$ and $\bar{\bar{\mathcal{G}}}_k$ are unnecessary to satisfy strong connection, thus Assumption 3.1 is very mild.

The objective of the consensus problem is to design an asynchronous impulsive protocol for all the followers, such that all agents can be synchronized to the leader agent for any initial conditions, i.e.,

$$\lim_{t\to\infty} \delta(t) = 0. \tag{3.5}$$

3.3 Main Results

3.3.1 Asynchronous Impulsive Controller Design

The distributed impulsive controller of agent i is designed as

$$u_i(t) = \sum_{k=1}^{\infty} b_k m_i(t_k) \left(\sum_{j\in N_i} a_{ij} \left(x_i(t) - x_j(t)\right) + c_i \left(x_i(t) - x_0(t)\right) \right) \delta^*(t - t_k), \tag{3.6}$$

where $k \in \mathbb{N}^+$, $i = 1, 2, \ldots, N$, b_k is the impulsive control gain. The impulsive sequence $\{t_k\}$ satisfying $0 \le t_0 < t_1 < t_2 < \cdots < t_{k-1} < t_k < \cdots$, $\lim_{k\to\infty} t_k = \infty$. $\delta(t)$ is the Dirac delta function, which satisfies $\delta(t) = 0$ for $t \neq 0$ and $\int_{-\infty}^{+\infty} \delta(t)\, dt = 1$. Define $\tau_k = t_k - t_{k-1}$ as the impulsive interval. At the impulsive instant $t = t_k$, if the agent i is controlled, then $m_i(t_k) = 1$, otherwise $m_i(t_k) = 0$.

From the model of agent i (3.1) and the impulsive control input (3.6), one gets the impulsive control system of the agent i is

$$\begin{cases} \dot{x}_i(t) = Ax_i(t) + f(x_i(t)), \quad t \neq t_k, \\ \Delta x_i(t_k) = x_i(t_k^+) - x_i(t_k^-) \\ \qquad = m_i(t_k) b_k \left(\sum_{j\in N_i} a_{ij} \left(x_i(t_k) - x_j(t_k)\right) + c_i \left(x_i(t_k) - x_0(t_k)\right) \right), \\ x_i(t_0^+) = x_i(t_0), \quad t_0 \ge 0, \quad i = 1, 2, \ldots N, \quad k \in \mathbb{N}, \end{cases} \tag{3.7}$$

where $\Delta x_i(t_k)$ is the jump of the state of node i at impulse instant t_k, and $x_i(t)$ is left continuous at t_k, i.e., $x_i(t_k) = x_i(t_k^-)$.

From Definition 3.1 and (3.7), one gets the global error system of agent i is

$$\begin{cases} \dot{\delta}_i(t) = A\delta_i(t) + f(x_i(t)) - f(x_0(t)), \quad t \neq t_k, \\ \Delta\delta_i(t_k) = m_i(t_k)b_k\left(\sum_{j\in N_i} a_{ij}\left(x_i(t_k) - x_j(t_k)\right) + c_i\left(x_i(t_k) - x_0(t_k)\right)\right), \end{cases} \tag{3.8}$$

where $1, 2, \ldots, N$.

From the definition of the Kronecker product, (3.8) can be rewritten as

$$\begin{cases} \dot{\delta}(t) = (I_N \otimes A)\delta(t) + \bar{f}(x(t), \bar{x}_0(t)), \quad t \neq t_k, \\ \Delta\delta(t_k) = M(t_k)b_k((L + C) \otimes I_n)\delta(t_k), \end{cases} \tag{3.9}$$

where $\bar{f}(x(t), \bar{x}_0(t)) = [f^T(x_1(t)) - f^T(x_0(t)), \ldots, f^T(x_N(t)) - f^T(x_0(t))]^T$, $x(t) = [x_1^T(t), x_2^T(t), \ldots, x_n^T(t)]^T$, $M(t_k) = \mathrm{diag}(m_1(t_k), m_2(t_k), \ldots, m_N(t_k))$, $\bar{x}_0(t) = 1_N \otimes x_0(t)$.

By using the controlled set, the error system of the agent $i \in Z(k)$ is

$$\begin{cases} \dot{\tilde{\delta}}_i(t) = A\tilde{\delta}_i(t) + f(x_i(t)) - f(x_0(t)), \quad t \neq t_k, \\ \Delta\tilde{\delta}_i(t_k) = b_k e_i(t_k). \end{cases} \tag{3.10}$$

The compact form of (3.10) is

$$\begin{cases} \dot{\tilde{\delta}}(t) = (I_N \otimes A)\tilde{\delta}(t) + \bar{f}(\tilde{x}(t), \tilde{\bar{x}}_0(t)), \quad t \neq t_k, \\ \Delta\tilde{\delta}(t_k) = b_k((\tilde{L}_k + \tilde{C}_k) \otimes I_n)\tilde{\delta}(t_k), \end{cases} \tag{3.11}$$

where $\tilde{\delta}(t) = [\delta^T_{z_1(k)}(t), \delta^T_{z_2(k)}(t), \ldots, \delta^T_{z_{p(k)}(k)}(t)]^T$, $\tilde{x}(t) = [x^T_{z_1(k)}(t), x^T_{z_2(k)}(t), \ldots, x^T_{z_{p(k)}(k)}(t)]^T$, $\bar{x}_0(t) = 1_{z_{p(k)}(k)} \otimes x_0(t)$, $\tilde{L}_k$ and $\tilde{C}_k$ is the Laplacian matrix and the leader adjacency matrix of $\tilde{\mathcal{G}}_k$.

3.3.2 Convergence Analysis

In this section, some inequality conditions are derived to achieve the consensus of nonlinear MASs via asynchronous impulsive protocol.

Notice that $\delta_i(t)$ cannot be measured directly if $c_i = 0$. In addition, for the agent i, if $c_i > 0$, and not be controlled when $t = t_k$, then $\delta_i(t_k)$ need not be measured because it is unnecessary to transmit information between node i and the leader. It should be noted that if the node i is not connected with the leader, i.e., $c_i = 0$, then the global synchronization error of agent i cannot be measured. In a word, when $t = t_k$, $V(t_k)$, $\tilde{V}(t_k)$, and $\bar{V}(t_k)$ does not need be measured, and only the agent i be controlled and $c_i > 0$, then $\delta_i(t_k)$ can be directly obtained. Construct a new set

$G(k) = \{g_1(k), g_2(k), \ldots, g_{q(k)}(k)\}$, and the agents in $G(k)$ are not only connected with the leader, but also controlled at the impulsive instant t_k.

Theorem 3.1 Assume the MASs satisfy Assumptions 1.1, 1.2, and 3.1, if

$$(\lambda_A + 2l)(t_{k+1} - t_k) + \ln(\eta_k \xi) \leq 0, \tag{3.12}$$

where $\xi > 1$, $\eta_k \in (\lambda_k, 1)$, $\lambda_k \in (0, 1)$ is the maximum eigenvalue of $\left(b_k(\tilde{L}_k + \tilde{C}_k) + I_N\right)^T \left(b_k(\tilde{L}_k + \tilde{C}_k) + I_N\right)$, then the consensus of MASs can be realized by the distributed asynchronous impulsive control designed in this chapter.

Proof Choose the Lyapunov function as

$$V(t) = \sum_{i=1}^{N} \delta_i^T(t)\delta_i(t) = \delta^T(t)\delta(t). \tag{3.13}$$

For $t \in (t_{k-1}, t_k]$,

$$\begin{aligned} D^+V(t) &= \dot{\delta}^T(t)\delta(t) + \delta^T(t)\dot{\delta}(t) \\ &= \delta^T(t)(I_N \otimes (A + A^T))\delta(t) + 2\delta^T(t)\bar{\psi}(x, \bar{x}_0) \\ &\leq (\lambda_A + 2l)\delta^T(t)\delta(t) \\ &= (\lambda_A + 2l)V(t), \end{aligned} \tag{3.14}$$

which leads to

$$V(t) \leq V((t_{k-1}^+)\exp((\lambda_A + 2l)(t - t_{k-1})). \tag{3.15}$$

When $t = t_k$,

$$\begin{aligned} &\tilde{V}(t_k^+) \\ &= \tilde{\delta}^T(t_k^+)\tilde{\delta}(t_k^+) \\ &= \left(\left(b_k\left((\tilde{L}_k + \tilde{C}_k) \otimes I_n\right) + I_{n \times p(k)}\right)\tilde{\delta}(t_k)\right)^T \\ &\quad \left(\left(b_k\left((\tilde{L}_k + \tilde{C}_k) \otimes I_n\right) + I_{n \times p(k)}\right)\tilde{\delta}(t_k)\right) \\ &= \tilde{\delta}^T(t_k)\left(\left(\left(b_k(\tilde{L}_k + \tilde{C}_k) + I_{p(k)}\right)^T \left(b_k(\tilde{L}_k + \tilde{C}_k) + I_{p(k)}\right)\right) \otimes I_n\right)\tilde{\delta}(t_k) \\ &\leq \lambda_k \tilde{V}(t_k), \end{aligned} \tag{3.16}$$

where $\tilde{V}(t_k^+) = \delta_{z_1(k)}^T(t_k)\delta_{z_1(k)} + \cdots + \delta_{z_p(k)}^T(t_k)\delta_{z_p(k)}$.

Denote $\bar{V}(t_k^+) = V(t_k^+) - \tilde{V}(t_k^+)$. Because the state of uncontrolled nodes does not jump at $t = t_k$, then one gets $\bar{V}(t_k^+) = \bar{V}(t_k)$, and

$$\begin{aligned} V(t_k^+) &= \tilde{V}(t_k^+) + \bar{V}(t_k^+) \\ &= \tilde{V}(t_k^+) + \bar{V}(t_k) \\ &\le \lambda_k \tilde{V}(t_k^+) + \bar{V}(t_k) \\ &\le \eta_k V(t_k). \end{aligned} \tag{3.17}$$

It should be noted that to apply (3.12) in Theorem 3.1, just knowing that $\eta_k \in (\lambda_k, 1)$ is not enough, it is necessary to determine the specific value of η_k. The following analysis is to determine the value of parameter η_k.

When $t \in (t_0, t_1]$, from (3.15), one gets

$$V(t) \le V(t_0) \exp((\lambda_A + 2l)(t - t_0)), \tag{3.18}$$

which leads to

$$V(t_1) \le V(t_0) \exp((\lambda_A + 2l)(t_1 - t_0)). \tag{3.19}$$

Define $\omega_k = V(t_k^+)/V(t_k)$, $\alpha_k = \tilde{V}_1(t_k)/\varphi(t_k)$, and $\varphi(t_1) = V(t_0)\exp((\lambda_A + 2l)(t_1 - t_0))$, $\varphi(t_k) = \eta_1 \cdots \eta_{k-1} V(t_0) \exp((\lambda_A + 2l)(t_k - t_0))$, $k = 2, 3, \ldots, N$, from (3.19), one gets $V(t_1) \le \varphi(t_1)$, then from the definition of α_k, one obtains

$$\alpha_1 = \tilde{V}_1(t_1)/\varphi(t_1) \le \tilde{V}_1(t_1)/V(t_1). \tag{3.20}$$

When $t = t_1$, from (3.17), one gets

$$\begin{aligned} \omega_1 &= V(t_1^+)/V(t_1) \\ &= (\tilde{V}(t_1^+) + \bar{V}(t_1^+))/V(t_1) \\ &\le (\lambda_1 \tilde{V}(t_1) + \bar{V}(t_1))/V(t_1) \\ &= (\lambda_1 \tilde{V}_1(t_1) + \lambda_1 \tilde{V}_2(t_1) + \bar{V}(t_1))/V(t_1) \\ &\le (\lambda_1 \tilde{V}_1(t_1) + \tilde{V}_2(t_1) + \bar{V}(t_1))/V(t_1), \end{aligned} \tag{3.21}$$

where $\tilde{V}_1(t_k) = \delta_{g_1(k)}^T(t_k)\delta_{g_1(k)}(t_k) + \cdots + \delta_{g_{q(k)}(k)}^T(t_k)\delta_{g_{q(k)}(k)}(t_k)$. Denote $\tilde{V}_2(t_k) = \tilde{V}(t_k) - \tilde{V}_1(t_k)$, from $\alpha_1 \le \tilde{V}_1(t_1)/V(t_1) \in (0, 1)$, one obtains

$$\tilde{V}_1(t_1) \ge \alpha_1 V(t_1) = \alpha_1(\tilde{V}_1(t_1) + \tilde{V}_2(t_1) + \bar{V}(t_1)), \tag{3.22}$$

which leads to

$$(1 - \alpha_1)\tilde{V}_1(t_1) \ge \alpha_1(\tilde{V}_2(t_1) + \bar{V}(t_1)). \tag{3.23}$$

Define $\beta_k = \alpha_k/(1-\alpha_k) > 0$, from (3.23), one gets

$$\tilde{V}_1(t_1) \geq \beta_1(\tilde{V}_2(t_1) + \bar{V}(t_1)). \tag{3.24}$$

From (3.21) and (3.24), one obtains

$$\begin{aligned} \omega_1 &\leq \frac{\lambda_1\beta_1(\tilde{V}_2(t_1) + \bar{V}(t_1)) + (\tilde{V}_2(t_1) + \bar{V}(t_1))}{\beta_1(\tilde{V}_2(t_1) + \bar{V}(t_1)) + (\tilde{V}_2(t_1) + \bar{V}(t_1))} \\ &= \frac{\lambda_1\beta_1 + 1}{\beta_1 + 1}. \end{aligned} \tag{3.25}$$

Denote $\eta_k = (\lambda_k\beta_k + 1)/(\beta_k + 1)$, then $V(t_1^+) = \omega_1 V(t_1) \leq \eta_1 V(t_1)$. From (3.17) and (3.19), one has

$$V(t_1^+) \leq \eta_1 V(t_1) \leq \eta_1 V(t_0)\exp((\lambda_A + 2l)(t_1 - t_0)). \tag{3.26}$$

When $t \in (t_1, t_2]$, from (3.15) and (3.26), one gets

$$\begin{aligned} V(t) &\leq V(t_1^+)\exp((\lambda_A + 2l)(t - t_1)) \\ &\leq \eta_1 V(t_0)\exp((\lambda_A + 2l)(t_1 - t_0))\exp((\lambda_A + 2l)(t - t_1)) \\ &\leq \eta_1 V(t_0)\exp((\lambda_A + 2l)(t - t_0)), \end{aligned} \tag{3.27}$$

which leads to

$$V(t_2) \leq \eta_1 V(t_0)\exp((\lambda_A + 2l)(t_2 - t_0)) = \varphi(t_2). \tag{3.28}$$

Since $\alpha_2 = \tilde{V}_1(t_2)/\varphi(t_2)$, from (3.26), one gets $\alpha_2 \leq \tilde{V}_1(t_2)/V(t_2)$, and

$$\begin{aligned} \omega_2 &\leq \frac{\lambda_2\beta_2(\tilde{V}_2(t_2) + \bar{V}(t_2)) + (\tilde{V}_2(t_2) + \bar{V}(t_2))}{\beta_2(\tilde{V}_2(t_2) + \bar{V}(t_2)) + (\tilde{V}_2(t_2) + \bar{V}(t_2))} \\ &= \frac{\lambda_2\beta_2 + 1}{\beta_2 + 1}, \end{aligned} \tag{3.29}$$

then one has $V(t_2^+) \leq \eta_2 V(t_2)$.

In general, when $t = t_k$,

$$\begin{aligned} V(t_k) &\leq V(t_{k-1}^+)\exp((\lambda_A + 2l)(t_k - t_{k-1})) \\ &\leq \eta_1 \cdots \eta_{k-1} V(t_0)\exp((\lambda_A + 2l)(t_k - t_0)) \\ &= \varphi(t_k), \end{aligned} \tag{3.30}$$

then one gets

$$\begin{aligned}\omega_k &\le \frac{\lambda_k\beta_k(\tilde{V}_2(t_k)+\bar{V}(t_k))+(\tilde{V}_2(t_k)+\bar{V}(t_k))}{\beta_k(\tilde{V}_2(t_k)+\bar{V}(t_k))+(\tilde{V}_2(t_k)+\bar{V}(t_k))}\\ &=\frac{\lambda_k\beta_k+1}{\beta_k+1}.\end{aligned}\tag{3.31}$$

Since $\eta_k=(\lambda_k\beta_k+1)/(\beta_k+1)\in(0,1)$, it is not difficult to obtain $V(t_k^+)\le\eta_k V(t_k)$ and $\eta_k\in(\lambda_k,1)$.

When $t\in(t_k,t_{k+1}]$, one obtains

$$\begin{aligned}V(t)&\le\eta_1\exp((\lambda_A+2l)(t_2-t_1))\cdots\eta_k\exp((\lambda_A+2l)(t_{k+1}-t_k))\\ &\quad V(t_0)\exp((\lambda_A+2l)(t_1-t_0))\\ &\le\frac{1}{\xi^k}V(t_0)\exp((\lambda_A+2l)(t_1-t_0)).\end{aligned}\tag{3.32}$$

Since $V(t_0)$, (λ_A+2l), and (t_1-t_0) are bounded constants, and $1/\xi^k\to 0$ when $k\to\infty$, then $\lim_{t\to\infty}V(t)=0$, one obtains the conclusion that $\delta(t)$ globally asymptotically approaches to zero. Therefore, the consensus of MASs is achieved under the asynchronous impulsive control designed in this chapter. Then the proof of Theorem 3.1 is completed.

Remark 3.2 In Theorem 3.1, $\lambda_k\in(0,1)$ is the most basic necessary condition to achieve the consensus of the MASs. Because of $\upsilon\ge 0$, if $\lambda_k\ge 1$, then $\eta_k>1$, and $V(t)\to 0$ as $t\to\infty$ cannot be achieved. Therefore, when $\eta_k>1$, it is impossible to achieve asynchronous impulsive consensus of MASs. In addition, from (3.12) in Theorem 3.1, one obtains the selectable range of impulsive interval is

$$t_{k+1}-t_k\le\frac{-\ln(\eta_k\xi)}{\lambda_A+2l}.\tag{3.33}$$

Due to the global consensus error of MASs will increase when there is no control input in continuous process, i.e., $(\lambda_A+2l)>0$. Therefore, when $\eta_k>1$, one gets $-\ln(\eta_k\xi)/(\lambda_A+2l)\le 0$, and then $t_{k+1}-t_k\le 0$, which is impossible in practice. In fact, if $\lambda_k\in(0,1)$, then $(\lambda_k\beta_k+1)<(\beta_k+1)$, which leads to $\eta_k\in(0,1)$. Based on the condition $\lambda_k\in(0,1)$, one can select the appropriate impulsive control gain b_k.

Remark 3.3 In Theorem 3.1, the variable ξ represents the convergence rate. From (3.33), one knows if the variable ξ increases, the smaller impulsive instants are obtained, and the convergence of the global synchronization error becomes faster, but the control cost will increase. When ξ is too large, it may lead to $\eta_k\xi>1$, resulting in a negative impulsive interval calculated according to (3.33), which is not feasible in practice. Therefore, in order to ensure the consensus of MASs under asynchronous impulsive control and save control consumption as much as possible, a parameter ξ slightly greater than 1 is selected.

Remark 3.4 Theorem 3.1 obtains the upper bound of two adjacent impulsive intervals, and the selection of impulsive interval is permissible as long as it is less than the upper bound. Because each impulsive control can play a role in promoting consensus, the global consensus error of the system can be reduced faster if the impulsive interval is small in the allowed range. However, if the impulsive interval is small, it will lead to a large control consumption. Therefore, on the premise of ensuring the consensus of MASs, in order to minimize the consumption of impulsive control, the impulsive interval is generally selected as the allowable upper bound value.

Remark 3.5 When $t = t_k$, it follows from (3.9) that

$$\begin{aligned} V(t_k^+) &= \delta^T(t_k^+)\delta(t_k^+) \\ &= ((b_k M(t_k)((L+C)\otimes I_n) + I_{nN})\delta(t_k))^T \\ &\quad ((b_k M(t_k)((L+C)\otimes I_n) + I_{nN})\delta(t_k)) \\ &= \delta^T(t_k)((b_k M(t_k)(L+C) + I_N)^T(b_k M(t_k)(L+C) + I_N)\otimes I_n)\delta(t_k) \\ &\le \bar{\lambda}_k V(t_k), \end{aligned}$$

where $\bar{\lambda}_k$ is the maximum eigenvalue of $(b_k M(t_k)(L+C)+I_N)^T(b_k M(t_k)(L+C)+I_N)$. When the asynchronous impulsive control is used, the diagonal matrix $M(t_k)$ contains element 0, which implies that 1 is an eigenvalue of $(b_k M(t_k)(L+C)+I_N)^T(b_k M(t_k)(L+C)+I_N)$, so it is impossible for $\bar{\lambda}_k$ to be less than 1. From (3.33), if $\bar{\lambda}_k \ge 1$, the obtained impulsive interval $t_{k+1} - t_k$ is negative, which is impossible to achieve in practice. Therefore, in the analysis of impulsive instants in this section, the representation in (3.9) is not used, and the diagonal matrix $M(t_k)$ also is not adopted. Instead, a new analysis method is adopted to obtain the ratio of the Lyapunov function (sum of squared global consensus errors of all nodes) before and after the impulsive instant.

3.3.3 Two Improved Shames

In Theorem 3.1, one can choose $\eta_k = (\lambda_k\beta_k + 1)/(\beta_k + 1)$ because β_k and λ_k can be obtained without the unmeasurable global consensus error. However, when one calculates the value of α_k, the estimated largest value of Lyapunov function $\varphi(t_k)$ is larger than the actual value $V(t_k)$, which increases the conservativeness of the result. To reduce the conservativeness of $\varphi(t_k)$, there are two schemes to be discussed.

Scheme 1

In the previous subsection, we indicate $\delta_i(t)$ cannot be measured directly if $c_i = 0$. However, in fact, if Assumption 3.1 holds and the agent i is controlled at $t = t_k$, one can calculate $\delta_i(t_k)$ by using global consensus error and local neighborhood synchronization error in MASs. For example, considering the topology $\bar{\mathcal{G}}$ in Fig. 3.1,

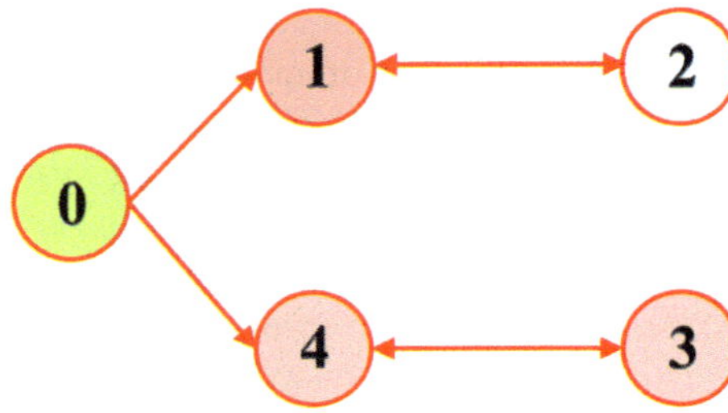

Fig. 3.1 Topology graph $\mathcal{G}$ of the MASs

to calculate the global consensus error of the agent 2, one could use the global consensus error of the agent 1 and the local neighborhood error of the agent 2, i.e.,

$$\begin{aligned}\delta_2(t) &= x_2(t) - x_0(t)\\ &= x_2(t) - x_1(t) + x_1(t) - x_0(t)\\ &= e_2(t) + \delta_1(t).\end{aligned} \tag{3.34}$$

Therefore, one can use $\tilde{V}(t_k)$ to estimate the value of η_k, which further reduces the conservativeness of the protocol in Theorem 3.1.

Let $\hat{\alpha}_k = \tilde{V}(t_k)/\varphi(t_k)$ and $\hat{\beta}_k = \hat{\alpha}_k/(1-\hat{\alpha}_k)$, when $t = t_1$, one obtains

$$\begin{aligned}\hat{\omega}_1 &= V(t_1^+)/V(t_1)\\ &= (\tilde{V}(t_1^+) + \bar{V}(t_1^+))/V(t_1)\\ &\leq (\lambda_1\tilde{V}(t_1) + \bar{V}(t_1))/V(t_1).\end{aligned} \tag{3.35}$$

Because of $V(t_1) \leq \bar{\varphi}(t_1)$, one gets $\hat{\alpha}_1 \leq \tilde{V}(t_1)/V(t_1)$, then

$$\tilde{V}_1(t_1) \geq \hat{\alpha}_1 V(t_1) \geq \hat{\alpha}_1(\tilde{V}(t_1) + \bar{V}(t_1)), \tag{3.36}$$

then one gets

$$\hat{\omega}_1 \leq \frac{\lambda_1\hat{\beta}_1\bar{V}(t_1) + \bar{V}(t_1)}{\hat{\beta}_1\bar{V}(t_1) + \bar{V}(t_1)} = \frac{\lambda_1\hat{\beta}_1 + 1}{\hat{\beta}_1 + 1}, \tag{3.37}$$

which implies $V(t_1^+) \leq \eta_1 V(t_1)$.

For any $t \in (t_k, t_{k+1}]$,

$$\hat{\omega}_k \leq \frac{\lambda_k\hat{\beta}_k\bar{V}(t_k) + \bar{V}(t_k)}{\hat{\beta}_k\bar{V}(t_k)) + \bar{V}(t_k)} = \frac{\lambda_k\hat{\beta}_k + 1}{\hat{\beta}_k + 1}. \tag{3.38}$$

Let $\eta_k = (\lambda_k\hat{\beta}_k + 1)/(\hat{\beta}_k + 1) \in (0, 1)$, then it is easy to obtain $V(t_k^+) \leq \eta_k V(t_k)$ and $\eta_k \in (\lambda_k, 1)$.

Remark 3.6 Scheme 1 reduces conservativeness of Theorem 3.1 to a certain extent. However, as time goes on, the conservativeness of $\varphi(t_k)$ becomes larger and larger and further causes the impulsive interval smaller and smaller. In practice, if the impulsive interval is too small, the impulsive control will become difficult to implement, and tiny impulsive intervals should be avoided in the design process of impulsive controller. Therefore, we design another scheme to reduce the conservatism of the asynchronous impulsive control strategy and avoid the situation that the impulsive interval is too small.

Scheme 2

In order to avoid $\varphi(t_k)$ becoming more and more conservative, one can measure the states of all agents in the MASs when $k = \vartheta\iota$ ($\vartheta, \iota \in \mathbb{N}$), and calculate the global synchronization error at $t = t_{\vartheta\iota}$, then one can get the actual value of $\varphi(t_k)$ when $k = \vartheta\iota$. Therefore, at $t = t_{\vartheta\iota}$, one can directly use the $\varphi(t_{\vartheta\iota})$ to eliminate the conservativeness. At other impulsive instants, one can use the method in Theorem 3.1 or Scheme 1 to obtain the value of $\varphi(t_k)$. In Scheme 2, the conservativeness of $\varphi(t_k)$ will be eliminated at $k = \vartheta, k = 2\vartheta, k = 3\vartheta \ldots$, which avoids the tiny impulsive interval.

Remark 3.7 The most difficult problem in this chapter is to find the appropriate value of η_k. From Theorem 3.1, Scheme 1 and Scheme 2, the obtained results have less conservativeness, but it needs more and more computation. Especially for Scheme 2, if the number of agents is too large, and the topological structure is complex, then it is difficult to calculate the global consensus error of all agents. Therefore, each method has its advantages and disadvantages, and how to choose the suitable method is based on the practical situation.

The selection of the estimated rules of variable η_k is given as follows:

(1) In order to avoid the large computation of global consensus error and local neighborhood error, and the conservativeness is within a tolerable range, then the method in Theorem 3.1 is chosen to achieve the consensus of the MASs.

(2) If one cannot tolerate the large conservativeness in Theorem 3.1, and a large amount of computation is acceptable, then the methods in Scheme 1 and Scheme 2 can be considered to reduce the conservativeness.

Remark 3.8 It should be noted that the analysis of continuous processes and impulsive instants in this chapter is highly conservative, and even if two improved schemes are provided, the designed control strategy still has its limitations. According to the inequality conditions in Theorem 3.1, the impulsive interval obtained is extremely small. However, using Scheme 2 requires measuring the global consensus error of the system at a certain impulsive instant, which will add a lot of additional computational complexity. Therefore, how to reduce the conservatism of continuous processes and impulsive instants without increasing the computational complexity required for control strategies is worthy of further research in future work.

Remark 3.9 The asynchronous impulsive control strategy designed in this chapter solves the problem of cannot control all agents at the same instant, that is, only controlling some agents at one impulsive instant can achieve the consensus of MASs. However, from Theorem 3.1 and the discussion on the selection of impulsive interval in Remark 3.2, it can be seen that the asynchronous impulsive control strategy designed in this chapter requires a fixed instant for impulse, i.e., the impulsive interval is fixed as the upper bound value obtained from the equation. Therefore, the asynchronous impulsive control designed in this chapter cannot solve the limitation of cannot exert impulsive control at a fixed instant precisely in practical engineering, and further improvement is needed in the future.

3.4 Simulation Results

The system model used in this section is the same as in Chap. 2 and still follows Chua's circuit as shown follows:

$$\begin{cases} \dot{x}_{i1} = p_1(x_{i2} - x_{i1} - g(x_{i1})), \\ \dot{x}_{i2} = x_{i1} - x_{i2} + x_{i3}, \\ \dot{x}_{i3} = -p_{i2}x_{i2}, \end{cases} \tag{3.39}$$

where p_1 and p_2 are system parameters, the expression of nonlinear function $g(\cdot)$ is

$$g(x_{i1}) = m_2 x_{i1} + 0.5(m_1 - m_2)\left(|x_{i1} + 1| - |x_{i1} - 1|\right). \tag{3.40}$$

The considered topology diagram is shown in Fig. 3.2, where the MASs consist of four follower nodes and one leader node.

Example 3.1 When $k = 1, 3, \ldots, 2\kappa - 1$ ($\kappa \in \mathbb{N}$), the agents 1, 2, and 4 are controlled at $t = t_k$. When $k = 2, 4, \ldots, 2\kappa$, the agents 1, 3, and 4 are controlled. Note that agents 1 and 4 are connected with the leader, and they are controlled at each impulsive instant. To reduce the conservativeness, the method in Scheme 2 is used to estimate the value of η_k.

In this example, the Laplacian matrix and leader adjacency matrix are obtained as

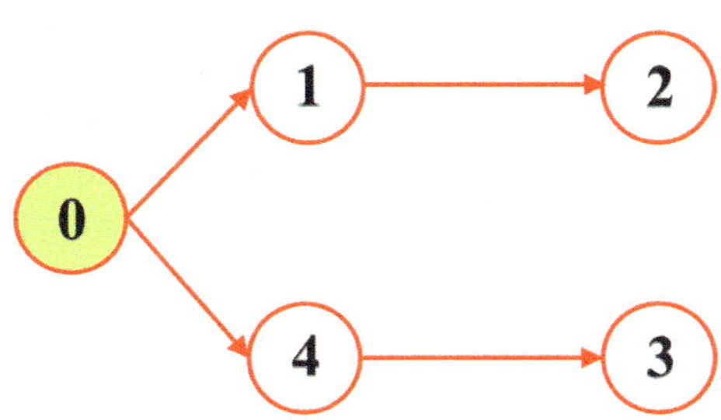

Fig. 3.2 Topology graph of the multi-agent systems

$$\begin{cases} \tilde{L}_k = \begin{bmatrix} 0 & 0 & 0 \\ -1 & 1 & 0 \\ 0 & 0 & 0 \end{bmatrix}, \; \tilde{C}_k = \begin{bmatrix} 1 & 0 & 0 \\ 0 & 0 & 0 \\ 0 & 0 & 1 \end{bmatrix}, \; k = 1, 3, \ldots, 2\kappa - 1, \\ \\ \tilde{L}_k = \begin{bmatrix} 0 & 0 & 0 \\ 0 & 1 & -1 \\ 0 & 0 & 0 \end{bmatrix}, \; \tilde{C}_k = \begin{bmatrix} 1 & 0 & 0 \\ 0 & 0 & 0 \\ 0 & 0 & 1 \end{bmatrix}, \; k = 2, 4, \ldots, 2\kappa. \end{cases}$$

According to the definition of parameters $\tilde{L}_k$, $\tilde{C}_k$, and λ_k, the relation between b_k and λ_k is presented in Fig. 3.3, then one knows when $b_k \in (-1, 0)$, $\lambda_k \in (0, 1)$. From the discussion in Remark 3.2, the selectable range of b_k is $b_k \in (-1, 0)$.

In Example 3.1, choose the initial value of each agent in MASs as $x_0 = [-2, 2, 3]^T$, $x_1 = [-3, 3, 1]^T$, $x_2 = [1, -1, -2]^T$, $x_3 = [2, -1, -1]^T$, $x_4 = [1, 0, -1]^T$. From Theorem 3.1, one calculates the allowable range of impulsive interval as

$$t_k - t_{k-1} \leq -\frac{\ln(\eta_k \xi)}{\lambda_A + 2l}. \tag{3.41}$$

Choose impulsive control gain $b_k = -0.5$, and the constant $\xi = 1.01$. Firstly, the analysis method in Theorem 3.1 is used to obtain the value of parameter η_k. According to Eq. (3.41), the relationship between impulsive interval and k can be obtained as shown in Fig. 3.4, where $\tau_k = t_k - t_{k-1}$.

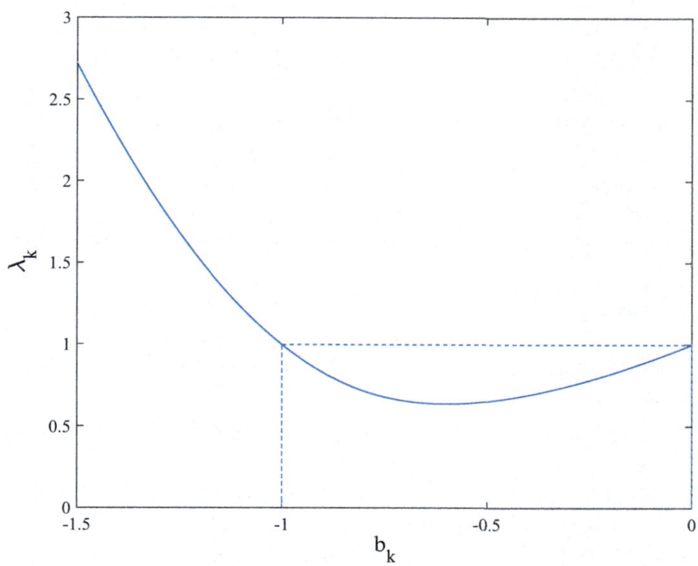

Fig. 3.3 The relation between impulsive control gain b_k and λ_k

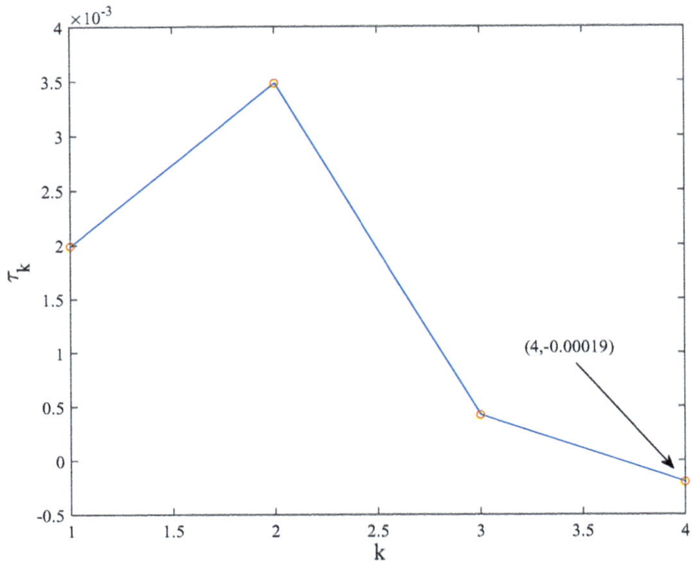

Fig. 3.4 τ_k and k obtained by Theorem 3.1 in Example 3.1

From Fig. 3.4, one knows the impulsive interval $\tau_k < 0$ when $k = 4$ via the method in Theorem 3.1, even though $\eta_k = 0.9948 \in (0, 1)$, $\eta_k \xi = 1.0047 > 1$, which leads to $\tau_k < 0$. This is not feasible in practice, so two improvement schemes are needed to avoid too small impulsive intervals. Consider Scheme 1 to obtain the value of parameter η_k, and then the relationship between impulsive interval and k obtained from Eq. (3.41) is shown in Fig. 3.5. It can be seen that the impulsive interval obtained by using Scheme 1 is slightly larger than that obtained directly using Theorem 3.1, but it is also negative at $k = 8$. At this point, although $\eta_k = 0.9971 \in (0, 1)$, due to $\eta_k \xi = 1.0047 > 1$, the obtained impulsive interval is negative, so using improved Scheme 1 is also not feasible. Therefore, in order to avoid the impulsive interval being too small or negative, only Scheme 2 can be used to obtain the value of η_k. Selecting parameter $\vartheta = 3$, which means that measure the state of all agents every 3 impulses and calculating the global consensus error of the MASs at the same time.

Figure 3.6 shows the relationship between the obtained impulsive interval when using Scheme 2. It can be seen that although the impulsive interval is not fixed, it has avoided the situation where the impulsive interval is very small or negative when using Theorem 3.1 or Scheme 1. Figure 3.7 shows the diagram of impulsive instants of the four following nodes. Different colors and symbols represent the impulsive instants of different agents. As shown in Fig. 3.7, at different impulsive instants, these four following agents are not controlled simultaneously, and the control law also meets the asynchronous impulse requirements specified in this example. In addition,

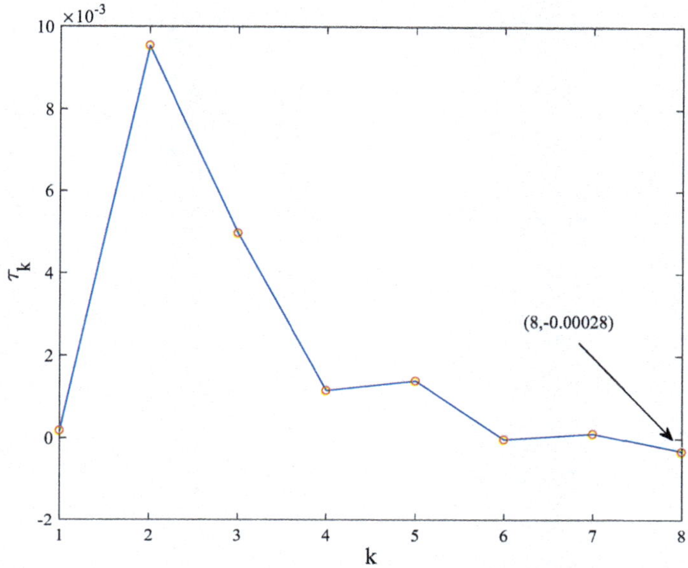

Fig. 3.5 τ_k and k obtained by Scheme 1 in Example 3.1

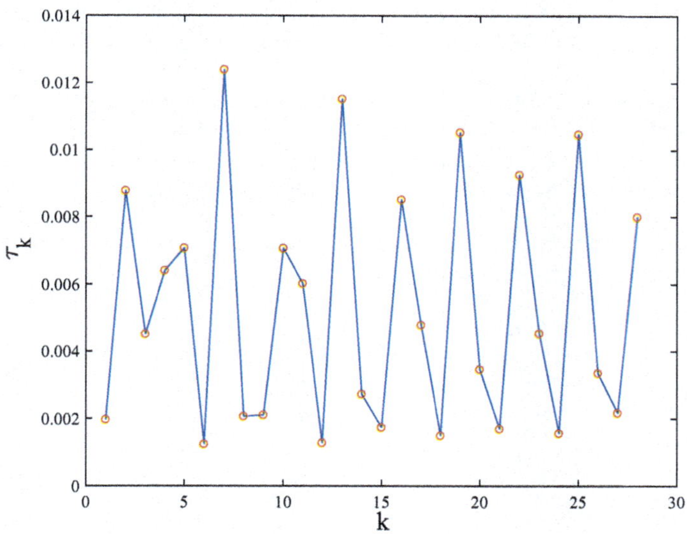

Fig. 3.6 τ_k and k obtained by Scheme 2 in Example 3.1

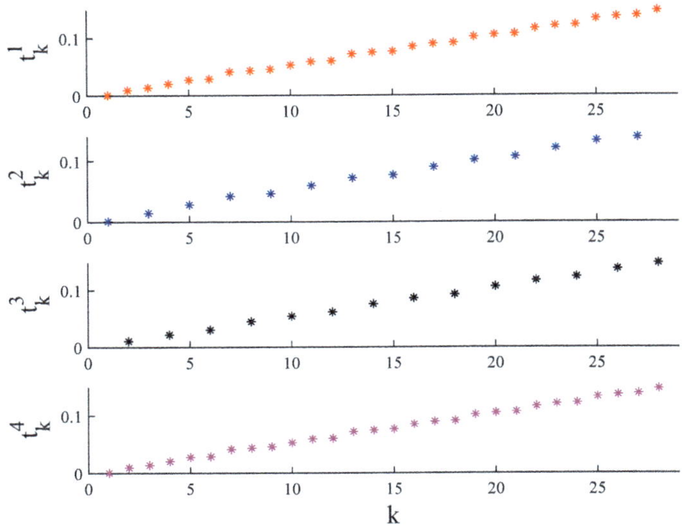

Fig. 3.7 The impulsive instants t_k^i of each agent in Example 3.1

as shown in Fig. 3.8, under the asynchronous impulsive control law designed in this chapter, the states of all following nodes can realize the consensus of the leader.

Example 3.2 In this example, when $k = 1, 3, \ldots, 2\kappa - 1$, the agents 1 and 2 are controlled at $t = t_k$, and when $k = 2, 4, \ldots, 2\kappa$, the agent 3 and 4 are controlled. Unlike Example 3.1, the asynchronous impulsive control strategy in this example does not have any agents to be controlled at all impulsive instants.

The initial values b_k and ξ are same as Example 3.1. The Laplacian matrix and leader adjacency matrix are obtained as

$$\begin{cases} \tilde{L}_k = \begin{bmatrix} 0 & 0 \\ -1 & 1 \end{bmatrix}, \ \tilde{C}_k = \begin{bmatrix} 1 & 0 \\ 0 & 0 \end{bmatrix}, \ k = 1, 3, \ldots, 2\kappa - 1 \\ \tilde{L}_k = \begin{bmatrix} 1 & -1 \\ 0 & 0 \end{bmatrix}, \ \tilde{C}_k = \begin{bmatrix} 0 & 0 \\ 0 & 1 \end{bmatrix}, \ k = 2, 4, \ldots, 2\kappa \end{cases}$$

Using Theorem 3.1 to calculate the value of η_k, then the curve of impulsive instant and k is presented in Fig. 3.9. When $k = 6$, $\tau_k < 0$. At this point, although $\eta_k = 0.9952 \in (0, 1)$, due to $\eta_k \xi = 1.005 > 1$, the obtained impulsive interval is negative, so it is necessary to use Scheme 1 or Scheme 2 to calculate the value of η_k.

Consider Scheme 1 to obtain η_k, and then the relationship between impulsive interval and k is shown in Fig. 3.10. It can be seen that the impulsive interval obtained by using Scheme 1 is negative at $k = 10$. At this point, although $\eta_k = 0.9944$, due to $\eta_k \xi = 1.004 > 1$, the obtained impulsive interval is negative, so using improved

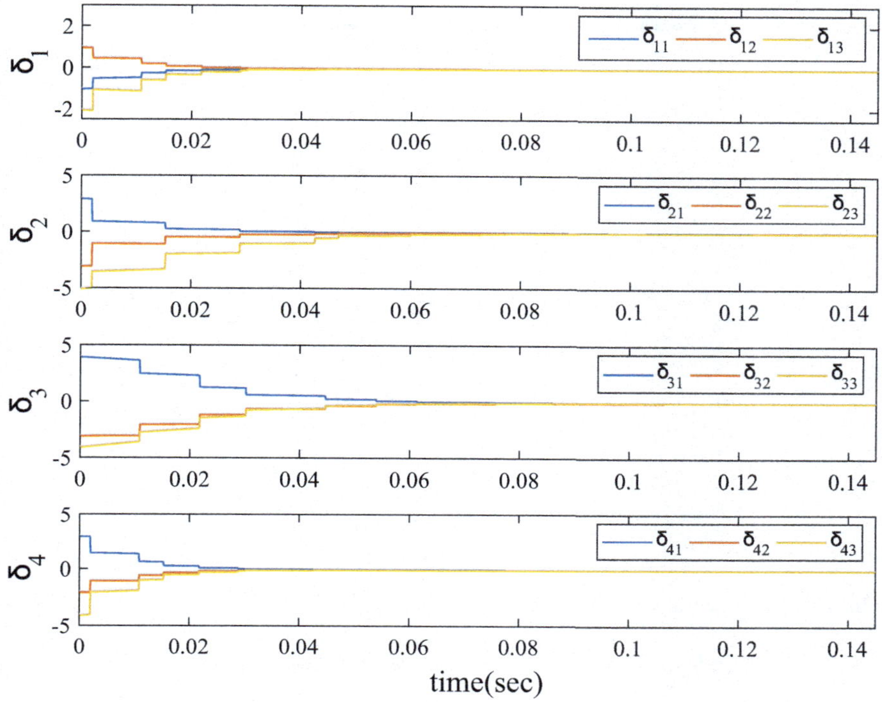

Fig. 3.8 The global consensus errors in Example 3.1

Scheme 1 is also not feasible. Therefore, only Scheme 2 can be used to obtain the value of η_k. Selecting parameter $\vartheta = 3$, the curve of impulsive interval is presented in Fig. 3.11.

From Fig. 3.11, it can be seen that using Scheme 2 can avoid situations where the impulsive interval is too small or negative. Figure 3.12 shows the impulsive instants of the four following agents, and it can be seen that the impulsive instants of the agents conform to the designed asynchronous impulsive control rules. In addition, Fig. 3.13 presents that the states of all following agents can achieve consensus with leader through the asynchronous impulsive control strategy designed in this chapter.

3.5 Summary

This chapter discusses the distributed asynchronous impulsive consensus protocol for nonlinear MASs. To find the appropriate value of η_k, there are three methods proposed in Theorem 3.1, Scheme 1 and Scheme 2. What's more, two simulations are shown to verify the correctness of the asynchronous impulsive schemes in this chapter.

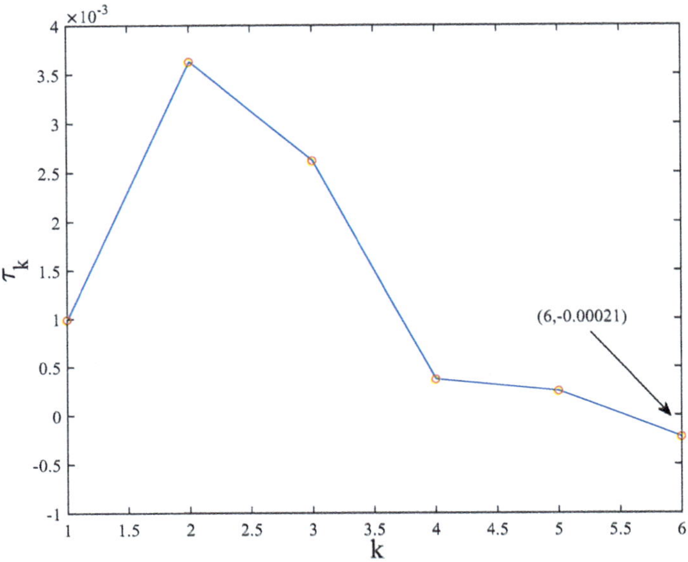

Fig. 3.9 τ_k and k obtained by Theorem 3.1 in Example 3.2

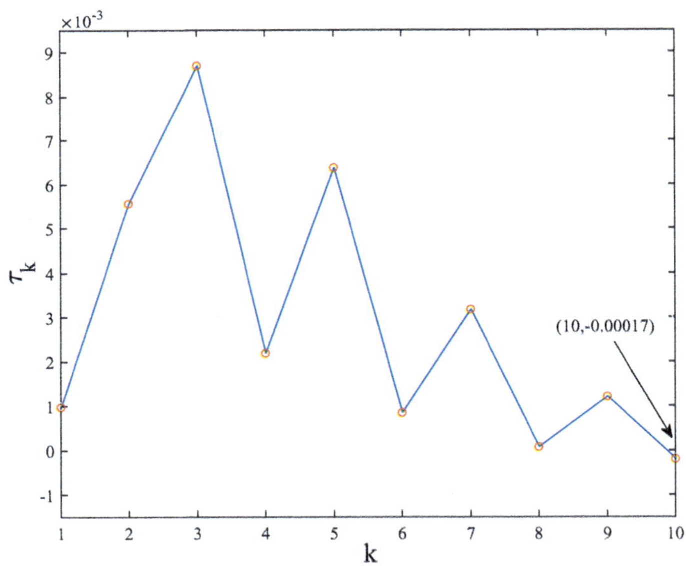

Fig. 3.10 τ_k and k obtained by Scheme 1 in Example 3.2

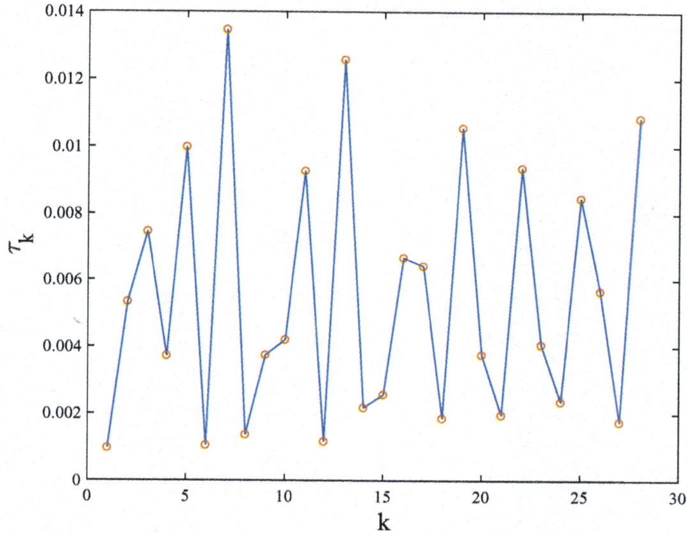

Fig. 3.11 τ_k and k obtained by Scheme 2 in Example 3.2

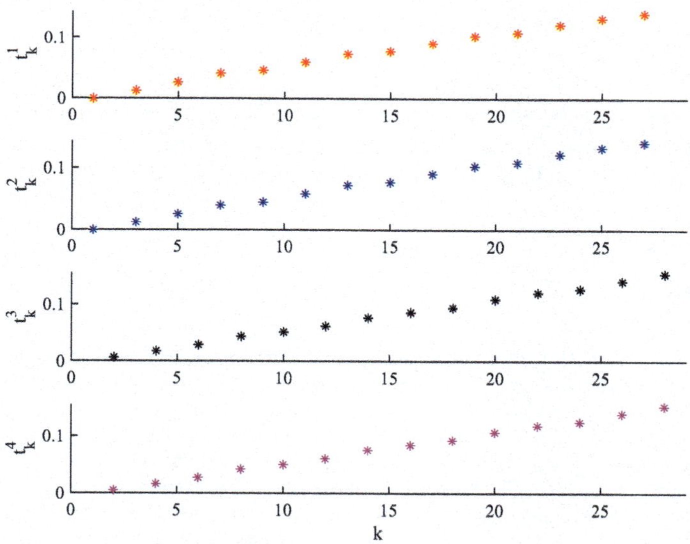

Fig. 3.12 The impulsive instants t_k^i of each agent in Example 3.2

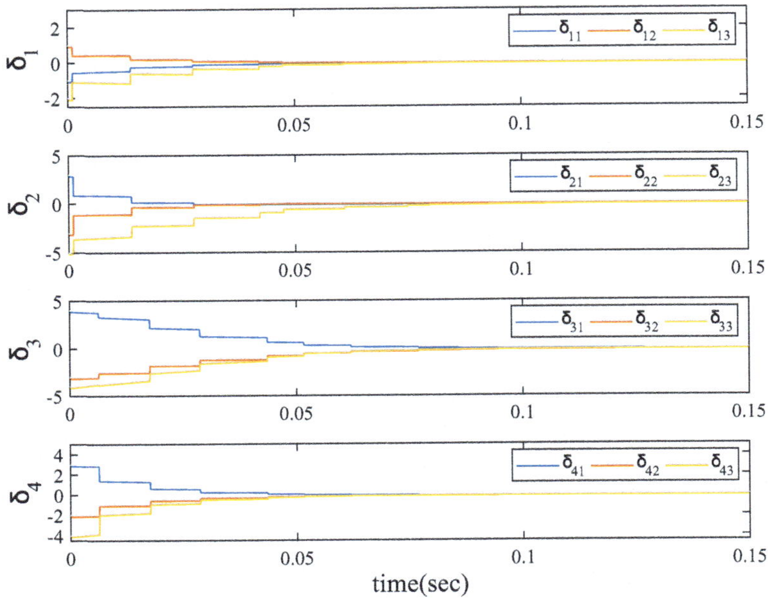

Fig. 3.13 The global consensus errors in Example 3.2

However, the analysis of the continuous process and the jump at impulsive instants still has conservativeness in this chapter. How to decrease the conservativeness of continuous process and impulsive instants deserves to be further explored.

References

Gong J, Ning D, Wu X, He G (2021) Bounded leader-following consensus of heterogeneous directed delayed multi-agent systems via asynchronous impulsive control. IEEE Trans Circuits Syst II Express Briefs 68(7):2680–2684

Han Y, Li C, Zeng Z (2019) Asynchronous event-based sampling data for impulsive protocol on consensus of non-linear multi-agent systems. Neural Netw 115:90–99

Jiang F, Liu B, Wu Y, Zhu Y (2018) Asynchronous consensus of second-order multi-agent systems with impulsive control and measurement time-delays. Neurocomputing 275:932–939

Li J, Zhang Y (2024) Asynchronous impulsive control for interval consensus of second-order multi-agent systems. Int J Control. https://doi.org/10.1080/00207179.2023.2301297

Liu ZW, Hu X, Ge MF, Wang YW (2019) Asynchronous impulsive control for consensus of second-order multi-agent networks. Commun Nonlinear Sci Numer Simul 79:104982

Zhang L, Li Y, Lou J, Lu J (2022) Bipartite asynchronous impulsive tracking consensus for multi-agent systems. Front Info Technol Electr Eng 23(10):1522–1532

Zhou B, Yang Y, Li L, Hao R (2022) Leaderless and leader-following consensus of heterogeneous second-order multi-agent systems on time scales: an asynchronous impulsive approach. Int J Control 95(9):2543–2553

Zhu W, Zhou Q, Li Q (2019) Asynchronous consensus of linear multi-agent systems with impulses effect. Commun Nonlinear Sci Numer Simul 82:105044

Chapter 4
Distributed Impulsive Consensus of Delayed Multi-agent Systems

4.1 Introduction

Due to the finite speed of transmitting and switching signals, the effects of time delay cannot be ignored in practical applications (Liu et al. 2017; Seuret et al. 2018; Grifa and Pepe 2021; Tian and Wang 2021). Recently, some research results for time delay have emerged in the coordination of MASs. Sun et al. (2023) addressed the issue of dynamic mean-squared consensus for second-order hybrid MASs, and the time-varying delays and multiplicative noises are considered. Fu and Liu (2023) considered the problem of containment control for Riemann–Liouville fractional multi-agent singular systems with time-varying delays. Zhuang et al. (2022) studied the leader-following exponential consensus problem of nonlinear stochastic discrete-time MASs, which are subject to parameter uncertainties, stochastic disturbances, and time-varying delay. Yu et al. (2022) investigated the sampled-based consensus problem of MASs with bounded distributed time delays and quantization effects. Xia et al. (2023) studied the containment control for fractional-order MASs with nonlinearity and time delay. Zhang et al. (2022a) investigated the sampled-data-based consensus control problem of MASs with multiplicative noise and time delays. It should be noted that the methods used in the above papers are continuous, which needs to continuously apply control to the MASs, resulting in a lot of control losses. Therefore, it is significant to investigate the consensus of delayed MASs by using the impulsive control method.

Recently, there are some results about impulsive consensus of MAS with time delay. Ke et al. (2020) focused on the distributed adaptive fault-tolerant consensus problem for nonlinear MASs using state-constraint impulsive protocols with time delay. Wei et al. (2022) investigated the impulsive average-consensus problem of first-order MASs with dynamically changing topologies. Zhu et al. (2022) studied the resilient group consensus of continuous-time second-order MASs with malicious agents. Xie et al. (2023) explored the event-triggered impulsive control for consensus problems in MASs with actuation delay. Hong and Zhang (2022)

Z. Zhang et al., *Distributed Impulsive Coordination of Multi-Agent Systems*, Intelligent Control and Learning Systems 15, https://doi.org/10.1007/978-981-97-6262-0_4

addressed the consensus problem of switched nonlinear MASs with delay and disturbances on undirected graph. You et al. (2024) considered the consensus problem of time-delayed fuzzy MASs with saturation-constraint impulsive control. Lyu et al. (2024) discussed the local consensus problem of nonlinear time-delayed MASs with switching topologies via distributed saturated impulsive control and estimated the maximum domain of attraction.

Practical systems are usually influenced by external disturbances. These disturbances are generally uncertain and therefore regarded as stochastic factors or stochastic effects (Li et al. 2021; Yang et al. 2021). In actual systems, stochastic disturbances are significant from theoretical and practical perspectives. Impulsive control with stochastic perturbations has been investigated previously. Chen et al. (2020) studied the impulsive pinning control of neutral stochastic delay Markov networks with stochastic disturbances. Wang et al. (2021) realized the stability of the impulsive system subject to stochastic disturbances via average dwell-time condition. Zhang et al. (2022b) realized the consensus of MASs subject to random nonlinearities by using a distributed impulsive protocol. Wang et al. (2019) studied the impulsive synchronization of stochastic delay complex networks. Liu et al. (2022) achieved the stabilization of a stochastic switching system through an event-triggered impulsive protocol.

Motivated by the aforementioned discussions, the stochastic synchronization of a class of nonlinear delayed MASs is addressed in this chapter. By combining the Lyapunov stability theory, impulsive differential equation, stochastic system, and mathematical induction approaches, a sufficient condition is established, which is more effective in practical systems.

4.2 Problem Description

The model of the leader is given by

$$\dot{x}_0(t) = Cx_0(t) + Af(x_0(t)) + Bf(x_0(t - \tau(t))), \tag{4.1}$$

where $x_0(t) \in \mathbb{R}^n$ is the state vector, $A, B, C \in \mathbb{R}^{n\times n}$ are the constant matrices. The transmission delay may be unknown but is bounded, i.e., $0 \leq \tau(t) \leq \tau$. The nonlinear function $f(x_0(t)) = [f_1(x_{01}(t)), \ldots, f_n(x_{0n}(t))]^T \in \mathbb{R}^n$.

Assumption 4.1 Each function $f_i : \mathbb{R} \to \mathbb{R}$ with $f_i(0) = 0$ satisfies the Lipschitz condition with a Lipschitz constant $\sigma_i > 0$, that is

$$|f_i(\varsigma_1) - f_i(\varsigma_2)| \leq \sigma_i|\varsigma_1 - \varsigma_2|, \forall \varsigma_1, \varsigma_2 \in \mathbb{R},\ i = 1, 2, \ldots, n.$$

Consider the controlled follower agents with stochastic disturbance. The dynamics of the ith $(i = 1, 2, \ldots, N)$ agent is given as

$$\begin{cases} dx_i(t) = (Cx_i(t) + Af(x_i(t)) + Bf(x_i(t-\tau(t))))dt \\ \qquad + \vartheta_i(t, e_i(t), e_i(t-\tau(t)))d\omega(t), \ t \neq t_k, \\ \Delta x_i(t_k) = x_i(t_k^+) - x_i(t_k^-) = x_i(t_k^+) - x_i(t_k) = u_i(t_k), \ k \in \mathbb{N}, \end{cases} \tag{4.2}$$

where $\{t_k\}$ satisfies $0 \leq t_0 < t_1 < t_2 \cdots < t_k < t_{k+1} < \cdots$, $\lim_{k\to\infty} t_k = \infty$, $\lim_{h\to 0^+} x_i(t_k + h) = x_i(t_k^+)$, $\lim_{h\to 0^+} x_i(t_k - h) = x_i(t_k^-) = x_i(t_k)$ implies that $x_i(t)$ is left continuous at $t = t_k$, $e_i(t) = x_i(t) - x_0(t)$ is the tracking error vector, $\omega(t)= [\omega_1(t), \ldots, \omega_n(t)]^T \in \mathbb{R}^n$ is the Brownian motion defined on $(\Omega, \mathcal{F}, \mathcal{F}_{t\geq 0}, \mathcal{P})$, and $\vartheta_i : \mathbb{R}^+ \times \mathbb{R}^n \times \mathbb{R}^n \to \mathbb{R}^{n\times n}$ is the diffusion coefficient matrix.

Assumption 4.2 $\vartheta_i(t, e_i(t), e_i(t-\tau(t)))$ is locally Lipschitz continuous and satisfies the linear growth condition, $\vartheta_i(t, 0, 0) = \mathbf{0}_n$. Moreover, there exist constant matrices of appropriate dimensions Γ_1 and Γ_2 such that

$$\begin{aligned} &\text{trace}[\vartheta_i^T(t, e_i(t), e_i(t-\tau(t)))\vartheta_i(t, e_i(t), e_i(t-\tau(t)))] \\ &\leq \|\Gamma_1 e_i(t)\|^2 + \|\Gamma_2 e_i(t-\tau(t))\|^2 . \end{aligned}$$

For a general stochastic nonlinear system $de(t) = g_1(t, e)dt + g_2(t, e)d\omega(t)$ on $t \geq t_0$ with initial value $e_0 \in \mathbb{R}^n$, $g_1 : \mathbb{R}^+ \times \mathbb{R}^n \to \mathbb{R}^n$ and $g_2 : \mathbb{R}^+ \times \mathbb{R}^n \to \mathbb{R}^{n\times n}$, for each $V(t, e) \in \mathcal{C}^{1,2}(\mathbb{R}^+ \times \mathbb{R}^n; \mathbb{R}^+)$, an operator $\mathcal{L}$ is defined from $\mathbb{R}^+ \times \mathbb{R}^n$ to $\mathbb{R}$ by

$$\mathcal{L}V(t, e) = V_t(t, e) + V_e(t, x)g_1(t, e) + \frac{1}{2}\text{trace}[g_2^T(t, e)V_{ee}(t, e)g_2(t, e)],$$

where

$$V_t(t, e) = \frac{\partial V(t,e)}{\partial t}, \ V_e(t, e) = \left(\frac{\partial V(t,e)}{\partial e_1}, \ldots, \frac{\partial V(t,e)}{\partial e_n}\right), \ V_{ee}(t, e) = \left(\frac{\partial V(t,e)}{\partial e_i \partial e_j}\right)_{n\times n}.$$

For simplicity, we denote $V_t(t, e)$, $V_e(t, e)$, and $V_{ee}(t, e)$ as $V_t(t)$, $V_e(t)$, and $V_{ee}(t)$, respectively.

From (4.1), (4.2), the fact $x_i(t_k^-) = x_i(t_k)$ and $x_0(t_k^+) = x_0(t_k^-) = x_0(t_k)$, the error system is obtained as

$$\begin{cases} de_i(t) = (Ce_i(t) + A(f(x_i(t)) - f(x_0(t))) \\ \qquad + B(f(x_i(t-\tau(t))) - f(x_0(t-\tau(t)))))dt \\ \qquad + \vartheta_i(t, e_i(t), e_i(t-\tau(t)))d\omega(t), \quad t \neq t_k, \\ \Delta e_i(t_k) = e_i(t_k^+) - e_i(t_k) = u_i(t_k), \ k \in \mathbb{N}, \\ e_i(t) = \varphi_i(t), \quad -\tau \leq t \leq t_0, \end{cases}$$

or the following compact form

$$\begin{cases} de(t) = ((I_N \otimes C)e(t) + (I_N \otimes A)h(e(t)) + (I_N \otimes B)h(e(t-\tau(t))))dt \\ \qquad +\vartheta(t, e(t), e(t-\tau(t)))d\varpi(t),\ t \neq t_k, \\ \Delta e(t_k) = e(t_k^+) - e(t_k) = U(t_k),\ k \in \mathbb{N}, \\ e(t) = \varphi(t), \quad -\tau \leq t \leq t_0, \end{cases} \tag{4.3}$$

where $e(t) = x(t) - \bar{x}_0(t) = [e_1^T(t), \ldots, e_N^T(t)]^T$, $x(t) = [x_1^T(t), \ldots, x_N^T(t)]^T$, $\bar{x}_0(t) = 1_N \otimes x_0(t)$, $\varpi(t) = 1_N \otimes \omega(t)$, $\varphi(t) = [\varphi_1^T(t), \ldots, \varphi_N^T(t)]^T$, $U(t) = [u_1^T(t), \ldots, u_N^T(t)]^T$, $\vartheta(t, e(t), e(t-\tau(t))) = \text{diag}\{\vartheta_i(t, e_i(t), e_i(t-\tau(t)))\}$, $h(e(t)) \doteq [(f(x_1(t)) - f(x_0(t)))^T, \ldots, (f(x_N(t)) - f(x_0(t)))^T]^T$, $h(e(t-\tau(t))) \doteq [(f(x_1(t-\tau(t))) - f(x_0(t-\tau(t))))^T, \ldots, (f(x_N(t-\tau(t))) - f(x_0(t-\tau(t))))^T]^T$.

The control objective is to design a distributed impulsive synchronization controller to let the states of the followers asymptotically synchronize with the leader in mean square, i.e., $\lim_{t\to\infty} E(\|e(t)\|^2) = 0$.

Before moving on, some lemmas are introduced, which will be used for the main results.

Lemma 4.1 *Suppose that positive continuous function $V(t)$ satisfies*

$$D^+V(t) \leq \hat{\alpha}V(t) + \hat{\beta}\tilde{V}(t), \quad \forall t \geq t_0, \tag{4.4}$$

then

$$E(V(t)) \leq E(\tilde{V}(t_0))e^{(\hat{\alpha}+\hat{\beta})(t-t_0)}, \quad \forall t \geq t_0,$$

where $\tilde{V}(t) \doteq \sup_{s\in[t-\tau,t]} V(s)$, $E(\tilde{V}(t)) \doteq \sup_{s\in[t-\tau,t]} E(V(s))$, $\hat{\alpha}$ and $\hat{\beta}$ are positive constants.

Proof The inequality (4.4) can be rewritten as

$$D^+V(t) \leq \hat{\alpha}V(t) + \hat{\beta}\tilde{V}(t) \leq (\hat{\alpha}+\hat{\beta})\tilde{V}(t), \quad \forall t \geq t_0. \tag{4.5}$$

Integrating both sides of (4.5) from t_0 to t, one gets

$$V(t) \leq V(t_0) + \int_{t_0}^{t} (\hat{\alpha}+\hat{\beta})\tilde{V}(\sigma)d\sigma.$$

Thus, for $s \in [t-\tau, t]$, one gets

$$V(s) \leq V(t_0) + \int_{t_0}^{s} (\hat{\alpha}+\hat{\beta})\tilde{V}(\sigma)d\sigma$$

and

$$
\begin{aligned}
\tilde{V}(t) &= \sup_{s\in[t-\tau,t]} V(s) \\
&\leq V(t_0) + \int_{t_0}^{\sup_{s\in[t-\tau,t]}\{s\}} (\hat{\alpha} + \hat{\beta})\tilde{V}(\sigma)d\sigma \\
&= V(t_0) + \int_{t_0}^{t} (\hat{\alpha} + \hat{\beta})\tilde{V}(\sigma)d\sigma,
\end{aligned}
$$

which implies

$$
V(t) \leq \tilde{V}(t) \leq V(t_0) + \int_{t_0}^{t} (\hat{\alpha} + \hat{\beta})\tilde{V}(\sigma)d\sigma.
$$

By using the Gronwall–Bellman inequality, we can get

$$
V(t) \leq \tilde{V}(t) \leq V(t_0)e^{(\hat{\alpha}+\hat{\beta})(t-t_0)} \leq \tilde{V}(t_0)e^{(\hat{\alpha}+\hat{\beta})(t-t_0)}, \forall t \geq t_0,
$$

which implies

$$
E(V(t)) \leq E(\tilde{V}(t_0))e^{(\hat{\alpha}+\hat{\beta})(t-t_0)}, \quad \forall t \geq t_0.
$$

This completes the proof.

Lemma 4.2 (Li et al. 2015) *Under Assumption 4.1, $L + G$ is nonsingular. Define*

$$
\begin{aligned}
&p = [p_1, \ldots, p_N]^T = ((L + G)^T)^{-1}1_N, \\
&P = diag\{p_i\}, \\
&Q = P(L + G) + (L + G)^T P.
\end{aligned}
\tag{4.6}
$$

Then $P > 0$ and $Q > 0$.

For simplicity, in the rest of this chapter, all the variables will be denoted without the time argument t, i.e., $x_i \doteq x_i(t)$, $e \doteq e(t)$, $x_{\tau i} \doteq x_i(t - \tau(t))$, $e_\tau \doteq e(t - \tau(t))$, and so on.

4.3 Main Results

4.3.1 Impulsive Controller Design

In this subsection, the consensus of delayed MASs with stochastic perturbation is realized.

To achieve the consensus goal, the following distributed tracking protocol is proposed.

$$
u_i(t_k) = b_{ik}\left(\sum_j a_{ij}(x_i(t_k) - x_j(t_k)) + g_i(x_i(t_k) - x_0(t_k))\right)
$$

or

$$U(t_k) = (B_k(L+G) \otimes I_n)e(t_k), \tag{4.7}$$

where $B_k = \text{diag}\{b_{ik}\} \in \mathbb{R}^{N \times N}$ is the control gain matrix to be designed. Pinning gain $g_i \geq 0$ is the weight of edges from the leader to agent i ($i \in \{1, \ldots, N\}$), where $g_i > 0$ if and only if there is an edge from the leader node to node i. Define $G = \text{diag}\{g_i\} \in \mathbb{R}^{N \times N}$.

From (4.3) and (4.7), one obtains the corresponding synchronization error dynamics described as

$$\begin{cases} de(t) = ((I_N \otimes C)e + (I_N \otimes A)h(e) \\ \qquad\qquad + (I_N \otimes B)h(e_\tau))dt + \vartheta(t, e, e_\tau)d\varpi, \ t \neq t_k, \\ \Delta e(t_k) = e(t_k^+) - e(t_k) = (B_k(L+G) \otimes I_n)e(t_k), \ k \in \mathbb{N}, \\ e(t) = \varphi(t), \quad -\tau \leq t \leq t_0. \end{cases} \tag{4.8}$$

4.3.2 Convergence Analysis

Theorem 4.1 *Assume that Assumptions 1.1, 4.1, and 4.2 hold. If there exist some positive constants α, β, $\eta_k < \xi < 1$ ($k \in \mathbb{N}$) and positive diagonal matrices Z_i ($i = 1, 2$) such that the following inequalities*

$$\begin{bmatrix} \Xi_i & I_N \otimes \Sigma \\ * & -Z_i \end{bmatrix} < 0, \tag{4.9}$$

$$\Upsilon_k < \lambda_{\min}(\bar{P})\eta_k I_{Nn}, \tag{4.10}$$

$$\left(\alpha + \frac{\beta}{\eta}\right)T + \ln \eta < \ln \xi \tag{4.11}$$

hold, where $\Xi_1 = 2P \otimes C + (P \otimes A)Z_1(P \otimes A)^T + (P \otimes B)Z_2(P \otimes B)^T + \lambda_{\max}(\bar{P})(I_N \otimes \Gamma_1^T \Gamma_1) - \lambda_{\min}(\bar{P})\alpha I_{Nn}$, $\Xi_2 = \lambda_{\max}(\bar{P})(I_N \otimes \Gamma_2^T \Gamma_2) - \lambda_{\min}(\bar{P}) \beta I_{Nn}$, $\Upsilon_k = ((B_k(L+G) \otimes I_n) + I_{Nn})^T \bar{P}((B_k \ (L+G) \otimes I_n) + I_{Nn})$, $\Sigma = \text{diag}\{\text{œ}_1, \ldots, \text{œ}_n\}$, $\bar{P} = P \otimes I_n$, *P is determined by Lemma 4.2.* $\eta = \sup\{\eta_k\}$, $T = \sup\{T_k\} = \sup\{t_{k+1} - t_k\}$, $k \in \mathbb{N}$. *Then the consensus of delayed MASs can be realized with controller* (4.7) *in mean square sense.*

Proof Choose the Lyapunov function as $V(t) = e^T \bar{P} e$, and it is easy to verify that $V(t)$ is nonnegative for $[t_0 - \tau, \infty)$. By Itô's differential formula, the stochastic derivative of $V(t)$ along the trajectory of system (4.8) is given as

$$dV(t) = \mathcal{L}V(t)dt + V_e(t)\vartheta(t, e, e_\tau)d\varpi. \tag{4.12}$$

For $t \in (t_k, t_{k+1}]$, $k \in \mathbb{N}$, taking the mathematical expectation on both sides of (4.12) and considering the fact of $E(V_e(t)\vartheta(t, e, e_\tau)d\varpi) = 0$, we get the following Dini derivative

$$E(D^+V(t)) = E(\mathcal{L}V(t)), \tag{4.13}$$

where

$$\begin{aligned}\mathcal{L}V(t) &= V_t(t) + V_e(t)((I_N \otimes C)e + (I_N \otimes A)h(e) \\ &\quad + (I_N \otimes B)h(e_\tau)) + \frac{1}{2}\text{trace}[\vartheta^T(t, e, e_\tau)V_{ee}(t, e)\vartheta(t, e, e_\tau)] \\ &= 2e^T\bar{P}((I_N \otimes C)e + (I_N \otimes A)h(e) \\ &\quad + (I_N \otimes B)h(e_\tau)) + \text{trace}[\vartheta^T(t, e, e_\tau)\bar{P}\vartheta(t, e, e_\tau)].\end{aligned} \tag{4.14}$$

The terms in (4.14) have the following properties:
(1) Using the property of the Kronecker product, it yields

$$2e^T\bar{P}(I_N \otimes C)e = 2e^T(P \otimes I_n)(I_N \otimes C)e = 2e^T(P \otimes C)e. \tag{4.15}$$

(2) Using Lemma 1.3 with $Z = Z_1$ and $\zeta = 1$, it yields

$$\begin{aligned}&2e^T\bar{P}(I_N \otimes A)h(e) \\ \le &e^T\bar{P}(I_N \otimes A)Z_1(I_N \otimes A)^T\bar{P}e + h^T(e)Z_1^{-1}h(e) \\ \le &e^T((P \otimes A)Z_1(P \otimes A)^T + (I_N \otimes \Sigma)Z_1^{-1}(I_N \otimes \Sigma))e.\end{aligned} \tag{4.16}$$

(3) Using Lemma 1.3 with $Z = Z_2$ and $\zeta = 1$, it yields

$$\begin{aligned}&2e^T\bar{P}(I_N \otimes B)h(e_\tau) \\ \le &e^T\bar{P}(I_N \otimes B)Z_2(I_N \otimes B)^T\bar{P}e + h^T(e_\tau)Z_2^{-1}h(e_\tau) \\ \le &e^T(P \otimes B)Z_2(P \otimes B)^Te + e_\tau^T(I_N \otimes \Sigma)Z_2^{-1}(I_N \otimes \Sigma)e_\tau.\end{aligned} \tag{4.17}$$

(4) Using Assumption 4.2, it yields

$$\begin{aligned}&\text{trace}[\vartheta^T(t, e, e_\tau)\bar{P}\vartheta(t, e, e_\tau)] \\ \le &\lambda_{\max}(\bar{P})(e^T(I_N \otimes \Gamma_1^T\Gamma_1)e + e_\tau^T(I_N \otimes \Gamma_2^T\Gamma_2)e_\tau).\end{aligned} \tag{4.18}$$

Thus, from (4.15) to (4.18), one has

$$\begin{aligned}
\mathcal{L}V(t) &\leq e^T(2P \otimes C + (P \otimes A)Z_1(P \otimes A)^T \\
&\quad + (I_N \otimes \Sigma)Z_1^{-1}(I_N \otimes \Sigma) + (P \otimes B)Z_2(P \otimes B)^T \\
&\quad + \lambda_{\max}(\bar{P})(I_N \otimes \Gamma_1^T \Gamma_1))e \\
&\quad + e_\tau^T((I_N \otimes \Sigma)Z_2^{-1}(I_N \otimes \Sigma) + \lambda_{\max}(\bar{P})(I_N \otimes \Gamma_2^T \Gamma_2))e_\tau \\
&= e^T(2P \otimes C + (P \otimes A)Z_1(P \otimes A)^T \\
&\quad + (I_N \otimes \Sigma)Z_1^{-1}(I_N \otimes \Sigma) + (P \otimes B)Z_2(P \otimes B)^T \\
&\quad + \lambda_{\max}(\bar{P})(I_N \otimes \Gamma_1^T \Gamma_1) - \alpha' I_{Nn})e \\
&\quad + e_\tau^T((I_N \otimes \Sigma)Z_2^{-1}(I_N \otimes \Sigma) \\
&\quad + \lambda_{\max}(\bar{P})(I_N \otimes \Gamma_2^T \Gamma_2) - \beta' I_{Nn})e_\tau + \alpha' e^T e + \beta' e_\tau^T e_\tau \\
&= e^T \Pi_1 e + e_\tau^T \Pi_2 e_\tau + \alpha' e^T e + \beta' e_\tau^T e_\tau,
\end{aligned}$$

where $\alpha' = \lambda_{\min}(\bar{P})\alpha$, $\beta' = \lambda_{\min}(\bar{P})\beta$, $\Pi_1 = 2P \otimes C + (P \otimes A)Z_1(P \otimes A)^T + (I_N \otimes \Sigma)Z_1^{-1}(I_N \otimes \Sigma) + (P \otimes B)Z_2(P \otimes B)^T + \lambda_{\max}(\bar{P})(I_N \otimes \Gamma_1^T \Gamma_1) - \lambda_{\min}(\bar{P})\alpha I_{Nn}$, $\Pi_2 = (I_N \otimes \Sigma)Z_2^{-1}(I_N \otimes \Sigma) + \lambda_{\max}(\bar{P})(I_N \otimes \Gamma_2^T \Gamma_2) - \lambda_{\min}(\bar{P})\beta I_{Nn}$.

By Schur complement Lemma, the condition (4.9) is equivalent to inequality $\Pi_i < 0$ $(i = 1, 2)$; therefore, one has

$$\begin{aligned}
\mathcal{L}V(t) &\leq \alpha' e^T e + \beta' e_\tau^T e_\tau \\
&\leq \frac{\alpha'}{\lambda_{\min}(\bar{P})} e^T \bar{P} e + \frac{\beta'}{\lambda_{\min}(\bar{P})} e_\tau^T \bar{P} e_\tau \\
&= \alpha V(t) + \beta V(t - \tau(t)).
\end{aligned} \tag{4.19}$$

From (4.13) and (4.19), we have

$$E(D^+ V(t)) \leq \alpha E(V(t)) + \beta E(V(t - \tau(t))). \tag{4.20}$$

When $t = t_k$, from the second and third equations of (4.8), it yields

$$e(t_k^+) = ((B_k(L + G) \otimes I_n) + I_{Nn})e(t_k),$$

and from (4.10), it further yields

$$\begin{aligned}
V(t_k^+) &= e^T(t_k^+)\bar{P}e(t_k^+) \\
&= e^T(t_k)((B_k(L + G) \otimes I_n) + I_{Nn})^T \\
&\quad \bar{P}((B_k(L + G) \otimes I_n) + I_{Nn})e(t_k) \\
&= e^T(t_k)\Upsilon_k e(t_k) \\
&\leq \lambda_{\min}(\bar{P})\eta_k e^T(t_k)e(t_k) \\
&\leq \eta_k V(t_k) \leq \eta V(t_k) \leq \eta \tilde{V}(t_k).
\end{aligned} \tag{4.21}$$

Taking the expectation on both sides of (4.21), it yields

$$E(V(t_k^+)) \le \eta E(V(t_k)) \le \eta E(\tilde{V}(t_k)),$$

where $E(\tilde{V}(t_k)) = \sup\limits_{s\in[t_k-\tau,t_k]} E(V(s))$.

Next, we will prove the following inequality holds,

$$V(t) < \xi\tilde{V}(t_k), \quad 0<\eta<\xi<1, \quad t \in (t_k, t_{k+1}], \quad k \in \mathbb{N}. \tag{4.22}$$

If (4.22) is not true, by using the continuity of $V(t)$ for $t \in (t_k, t_{k+1}]$ and $V(t_k^+) \le \eta\tilde{V}(t_k)$ (see (4.21)), there must exist a $\tilde{t}_k \in (t_k, t_{k+1}]$ such that

$$V(\tilde{t}_k) = \xi\tilde{V}(t_k) \tag{4.23}$$

and

$$V(t) < \xi\tilde{V}(t_k), \text{ for } t \in (t_k, \tilde{t}_k).$$

From (4.23) and $V(t_k^+) \le \eta\tilde{V}(t_k)$, there must exist $\hat{t}_k \in (t_k, \tilde{t}_k)$ such that

$$V(\hat{t}_k) = \eta\tilde{V}(t_k) \tag{4.24}$$

and

$$\eta\tilde{V}(t_k) \le V(t) \le \xi\tilde{V}(t_k), \text{ for } t \in [\hat{t}_k, \tilde{t}_k],$$

where $\hat{t}_k = \sup\{t \in (t_k, \tilde{t}_k), V(t) \le \eta\tilde{V}(t_k)\}$.

For $t \in [\hat{t}_k, \tilde{t}_k]$, one gets

$$V(t+s) \le \tilde{V}(t_k), \text{ for } s \in [-\tau, 0]. \tag{4.25}$$

It follows from (4.24) and (4.25) that

$$\eta V(t+s) \le \eta\tilde{V}(t_k) \le V(t), \text{ for } s \in [-\tau, 0), \ t \in [\hat{t}_k, \tilde{t}_k].$$

Thus, for $t \in [\hat{t}_k, \tilde{t}_k]$, (4.20) yields

$$E(D^+V(t)) \le \alpha E(V(t)) + \beta E(V(t-\tau(t))) \le \left(\alpha + \frac{\beta}{\eta}\right) E(V(t)). \tag{4.26}$$

Integrating both sides of (4.26) from $\hat{t}_k$ to $\tilde{t}_k$, where $t_k < \hat{t}_k < \tilde{t}_k < t_{k+1}$, it yields

$$\ln(E(V(\tilde{t}_k))) - \ln(E(V(\hat{t}_k))) \le \left(\alpha + \frac{\beta}{\eta}\right)(\tilde{t}_k - \hat{t}_k) \le \left(\alpha + \frac{\beta}{\eta}\right)T. \tag{4.27}$$

On the other hand, from (4.11), (4.23), and (4.24), one gets

$$\begin{aligned}\ln(E(V(\tilde{t}_k))) - \ln(E(V(\hat{t}))) &= \ln(\xi E(\tilde{V}(t_k))) - \ln(\eta E(\tilde{V}(t_k))) \\ &= \ln \xi - \ln \eta > \left(\alpha + \frac{\beta}{\eta}\right) T,\end{aligned} \tag{4.28}$$

which contradicts (4.27), and therefore (4.22) holds. Taking the expectation on both sides of (4.22), it yields

$$E(V(t)) < \xi E(\tilde{V}(t_k)), \; 0 < \eta < \xi < 1, \; t \in (t_k, t_{k+1}], \; k \in \mathbb{N}. \tag{4.29}$$

Next, we prove one basic fact $E(\tilde{V}(t_k)) \leq E(\tilde{V}(t_{k-1})), \quad k \in \mathbb{N}$.

$$\begin{aligned} E(\tilde{V}(t_k)) &= \sup_{s \in [t_k - \tau, t_k]} E(V(s)) \\ &= \max\{ \sup_{s \in [t_k - \tau, t_k)} E(V(s)), E(V(t_k^+))\} \\ &\leq \max\{ \sup_{s \in [t_k - \tau, t_k)} E(V(s)), \eta E(V(t_k^-))\} \\ &\leq \max\{ \sup_{s \in [t_k - \tau, t_k)} E(V(s)), E(V(t_k^-))\} \\ &= \sup_{s \in [t_k - \tau, t_k)} E(V(s)). \end{aligned} \tag{4.30}$$

(i) If $t_k - \tau \geq t_{k-1}$, then (4.30) yields

$$E(\tilde{V}(t_k)) \leq \sup_{s \in [t_k - \tau, t_k)} E(V(s)) \leq \xi E(\tilde{V}(t_{k-1})).$$

(ii) If $t_k - \tau < t_{k-1}$, then (4.30) yields

$$\begin{aligned} E(\tilde{V}(t_k)) &\leq \sup_{s \in [t_k - \tau, t_k)} E(V(s)) \\ &\leq \max\{E(\tilde{V}(t_{k-1})), \xi E(\tilde{V}(t_{k-1}))\} \\ &= E(\tilde{V}(t_{k-1})). \end{aligned}$$

From the above two cases, we have

$$E(\tilde{V}(t_k)) \leq E(\tilde{V}(t_{k-1})), \quad k \in \mathbb{N}. \tag{4.31}$$

Without loss of generality, let there exist $1 < k_1 < k_2 < \cdots < k_{(\chi-1)} < k_\chi < k_{(\chi+1)} < \cdots \quad (k_j \in \mathbb{N}, j \in \mathbb{N})$ such that

$$\begin{cases} t_{k_1} - \tau \in (t_1, t_2] \\ t_{k_2} - \tau \in (t_{k_1}, t_{k_1+1}], \\ \vdots \\ t_{k_{(\chi-1)}} - \tau \in (t_{k_{(\chi-2)}}, t_{k_{(\chi-2)}+1}], \\ t_{k_\chi} - \tau \in (t_{k_{(\chi-1)}}, t_{k_{(\chi-1)}+1}], \\ t_{k_{(\chi+1)}} - \tau \in (t_{k_\chi}, t_{k_\chi+1}], \\ \vdots \end{cases} \tag{4.32}$$

Thus, for $t \in (t_{k_\chi}, t_{k_{(\chi+1)}}]$, it follows from (4.29), (4.31), and (4.32) that

$$E(V(t)) \leq \xi E(\tilde{V}(t_{k_\chi})). \tag{4.33}$$

Furthermore, from (4.32), one gets $t_{k_{\chi+1}} - \tau > t_{k_\chi}$, which yields

$$E(\tilde{V}(t_{k_{\chi+1}})) \leq \xi E(\tilde{V}(t_{k_\chi})). \tag{4.34}$$

Thus, from (4.34), (4.33) can further yield that

$$\begin{aligned} E(V(t)) &\leq \xi E(\tilde{V}(t_{k_\chi})) \\ &\leq \xi^2 E(\tilde{V}(t_{k_{\chi-1}})) \\ &\leq \xi^3 E(\tilde{V}(t_{k_{\chi-2}})) \\ &\vdots \\ &\leq \xi^\chi E(\tilde{V}(t_{k_1})) \\ &\leq \xi^{\chi+1} E(\tilde{V}(t_1)). \end{aligned} \tag{4.35}$$

It follows from (4.31) that (4.35) further yields that

$$\begin{aligned} E(V(t)) &\leq \xi^{\chi+1} E(\tilde{V}(t_1)) \\ &\leq \xi^{\chi+1} E(\tilde{V}(t_0)), \end{aligned}$$

which implies that

$$E(\|e(t)\|^2) \leq \xi^{\chi+1} \frac{E(\tilde{V}(t_0))}{\lambda_{\min}(\bar{P})} = \xi^{\chi+1}\sigma. \tag{4.36}$$

Note that $\sigma \doteq \frac{E(\tilde{V}(t_0))}{\lambda_{\min}(\bar{P})}$ is finite positive constant, and $\xi^{\chi+1} \to 0$ as $\chi \to \infty$ (i.e., $t \to \infty$). Therefore one gets $\lim_{t\to\infty} E(\|e(t)\|^2) = 0$. This completes the proof.

Remark 4.1 Note that matrix P appears in matrix Ξ_i of (4.9), and these entries depend on the graph topology $L + G$ through Lemma 4.2. Therefore, for different graph topologies, the designs of the controller parameters T_k and B_k are also different. Moreover, the forms $\lambda_{\max}(\bar{P})$ and $\lambda_{\min}(\bar{P})$ appear in (4.9), and their values are related to $L+G$ through Lemma 4.2 (note that $\lambda_{\min}(\bar{P}) = \lambda_{\min}(P)$, $\lambda_{\max}(\bar{P}) = \lambda_{\max}(P)$, and P is determined by (4.6) of Lemma 4.2).

Remark 4.2 To guarantee the validity of the inequality (4.10), the topology parameter of multi-agent networks (correspond to $L+G$) is one of the important parameters. Thus, besides Assumption 1.1, the topology should be chosen as the suitable structure to satisfy the constraint condition in Theorem 4.1.

Remark 4.3 In this section, the control objective is to obtain the maximal impulsive interval T under fixed impulsive gain matrix B_k. From (4.11) of Theorem 4.1, T is determined by some elements α, β, and η, which further is determined by system structure parameters (matrices A, B, C, Σ, Γ_1, Γ_2), system topology ($L+G$), and impulsive gains B_k. How to obtain the maximal stable region with control parameters (T_k and B_k) is a challenging and significant problem.

Remark 4.4 By the recursive results of (4.22) and (4.31), the information of time delay $\tau(t)$ is included in $E(\tilde{V}(t_k))$. Finally, it follows from the recursive result (4.36) that the asymptotic convergence of error vector $e(t)$ is derived, where inequality (4.36) only includes $E(\tilde{V}(t_0))$ (one constant) instead of $E(\tilde{V}(t_k))$. Thus, Theorem 4.1 is delay-independent (synchronization conditions (4.9)–(4.11) do not contain the information of time delay). It is noteworthy that the time delay $\tau(t)$ still has some influence on the convergence performance. From (4.32), the value of τ affects the value of t_{k_χ}, or to be more specific, smaller τ corresponds to smaller t_{k_χ}. Here we suppose two different delay bounds $0 < \check{\tau} < \hat{\tau}$ correspond to $\check{t}_{k_\chi}$ and $\hat{t}_{k_\chi}$, respectively, and it still satisfies $0 < \check{t}_{k_\chi} < \hat{t}_{k_\chi}$. From inequality (4.36), one can get $E(\|e(t)\|^2) \leq \xi^{\chi+1}\sigma$ for $t \in (t_{k_\chi}, t_{k_{(\chi+1)}}]$. Obviously, the required time for $t \in (\check{t}_{k_\chi}, \check{t}_{k_{(\chi+1)}}]$ is smaller than the one for $t \in (\hat{t}_{k_\chi}, \hat{t}_{k_{(\chi+1)}}]$ although they have the same convergence rate (i.e., they have the same term $\xi^{\chi+1}\sigma$), and thus it is not hard to conclude smaller τ implies fast convergence speed.

4.4 Simulation Results

The dynamics of the followers and leader can be described as follows (Ma and Fu 2011):

$$\begin{cases} \dot{x}_i = Cx_i + Af(x_i) + Bf(x_{\tau i}), & t \neq t_k, \\ \Delta x_i(t_k) = u_i(t_k), \end{cases}$$

$$\dot{x}_0 = Cx_0 + Af(x_0) + Bf(x_{\tau 0}),$$

where $C = -I_2$, $f_i(x_{ij}) = (|x_{ij}+1| - |x_{ij}-1|)/2$, $i = 0, 1, 2, 3, 4$, $j = 1, 2$, $\tau(t) \equiv \tau = 1$. From Assumption 4.1, it is easy to verify that $\Sigma = I_2$. Let

$\vartheta_i(t, e_i, e_{\tau i})= diag\{1/\sqrt{2}(e_{i1} + e_{\tau i1}), 1/\sqrt{2}(e_{i2} + e_{\tau i2})\}$, hence we can easily get $\Gamma_1 = \Gamma_2 = I_2$ from Assumption 4.2. The constant matrices are specified as

$$A = \begin{bmatrix} 1+\pi/4 & 20 \\ 0.1 & 1+\pi/4 \end{bmatrix}, \quad B = \begin{bmatrix} -1.3\sqrt{2}\pi/4 & 0.1 \\ 0.1 & -1.3\sqrt{2}\pi/4 \end{bmatrix}.$$

Example 4.1 Consider the MASs with fixed topology $\mathcal{G}_1$ as shown in Fig. 4.1. From the graph theory, one gets

$$\mathcal{A} = \begin{bmatrix} 0&0&0&0 \\ 1&0&0&0 \\ 0&1&0&0 \\ 1&0&0&0 \end{bmatrix}, \mathcal{D} = \begin{bmatrix} 0&0&0&0 \\ 0&1&0&0 \\ 0&0&1&0 \\ 0&0&0&1 \end{bmatrix}$$

$$L = \mathcal{D} - \mathcal{A} = \begin{bmatrix} 0&0&0&0 \\ -1&1&0&0 \\ 0&-1&1&0 \\ -1&0&0&1 \end{bmatrix}, C = \begin{bmatrix} 1&0&0&0 \\ 0&0&0&0 \\ 0&0&0&0 \\ 0&0&0&0 \end{bmatrix}.$$

Select the parameters $\alpha = 164.42$, $\beta = 10.4$, $B_k = -0.73I_4$, $\eta_k = \eta = 0.42$, $\xi = 0.895$, $T_k = T = 0.004$, $\chi_\phi = 16$, $\Delta_\phi = 0.01$, $\tau_\phi - \tau_{\phi-1} = 0.074$s such that (4.9)–(4.11) hold, and the corresponding communication rate is $\mu_\phi = 0.86$. The system responses under the proposed impulsive control are shown in Figs. 4.2 and 4.3. From Figs. 4.2 and 4.3, one gets the consensus of time-delayed MASs with stochastic perturbation can be realized by using the impulsive control protocol designed in this chapter.

Example 4.2 In order to better illustrate the effectiveness and universality of the impulsive control protocol designed in this chapter, another topology structure $\mathcal{G}_1$ is considered in this example, which is presented in Fig. 4.4. From the graph theory, one gets

$$\mathcal{A} = \begin{bmatrix} 0&0&0&0 \\ 1&0&0&0 \\ 0&0&0&1 \\ 0&0&0&0 \end{bmatrix}, \mathcal{D} = \begin{bmatrix} 0&0&0&0 \\ 0&1&0&0 \\ 0&0&1&0 \\ 0&0&0&0 \end{bmatrix}$$

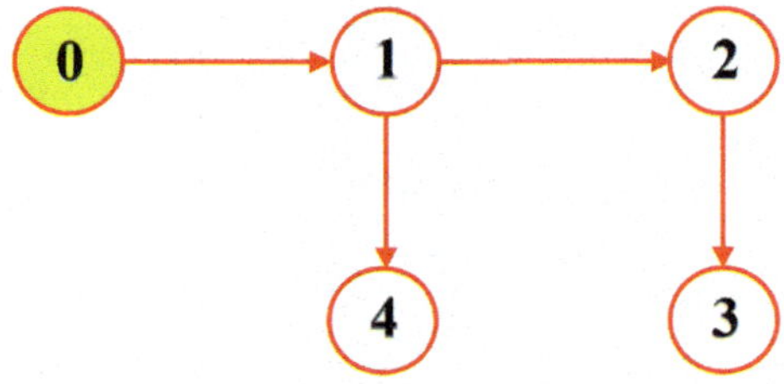

Fig. 4.1 Topology graph of the multi-agent systems in Example 4.1

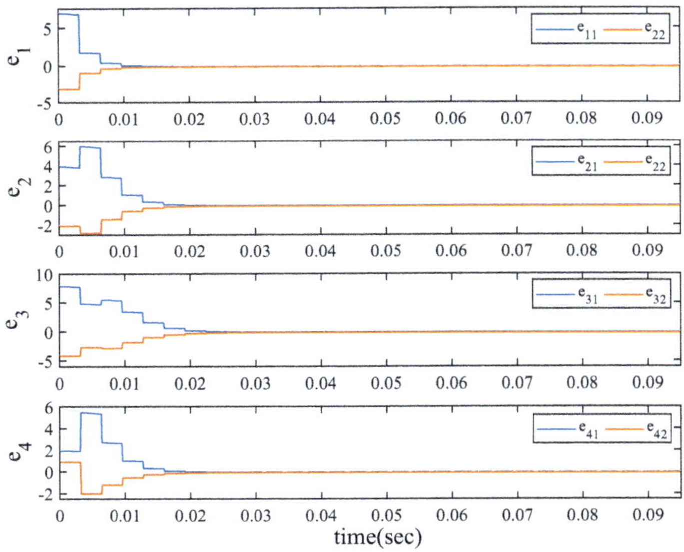

Fig. 4.2 The global consensus error $e(t)$ for all agents in Example 4.1

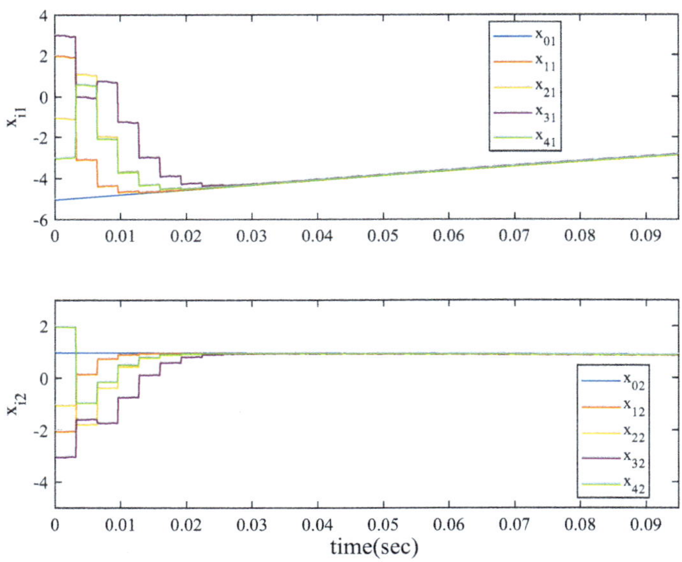

Fig. 4.3 State trajectories of $x_i (i = 0, 1, 2, 3, 4)$ for all agents in Example 4.1

$$L = \mathcal{D} - \mathcal{A} = \begin{bmatrix} 0 & 0 & 0 & 0 \\ -1 & 1 & 0 & 0 \\ 0 & 0 & 1 & -1 \\ 0 & 0 & 0 & 0 \end{bmatrix}, C = \begin{bmatrix} 1 & 0 & 0 & 0 \\ 0 & 0 & 0 & 0 \\ 0 & 0 & 0 & 0 \\ 0 & 0 & 0 & 1 \end{bmatrix}.$$

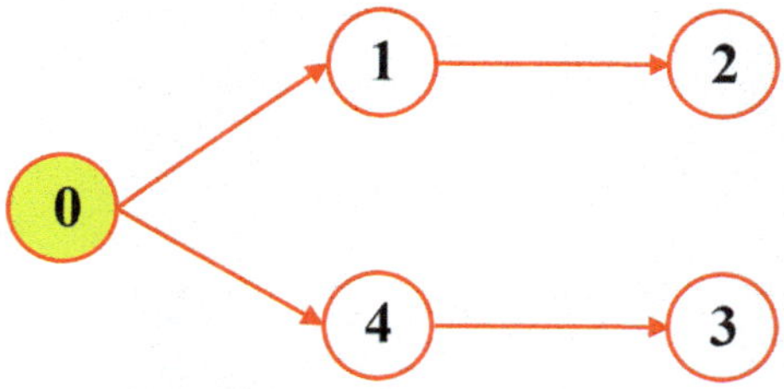

Fig. 4.4 Topology graph of the multi-agent systems in Example 4.2

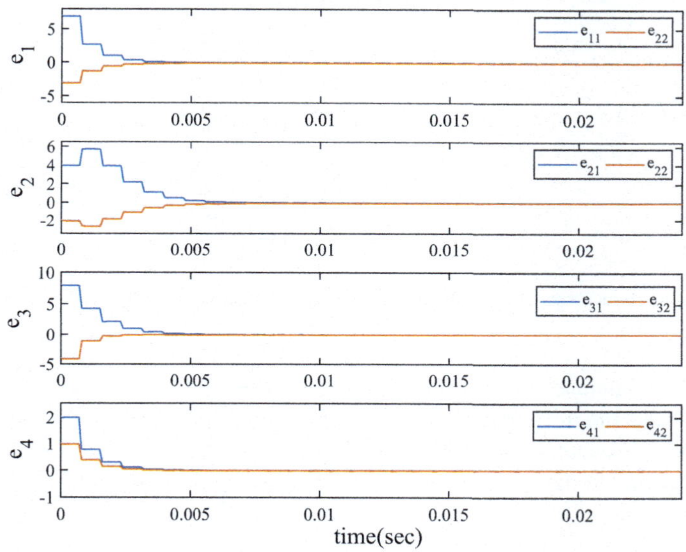

Fig. 4.5 The global consensus error $e(t)$ for all agents in Example 4.2

In this example, choose $B_k = -0.6I_4$, then $\eta_k = \eta = 0.7738$. From Theorem 4.1, we obtain $T = 0.0008$. Keeping other parameters unchanged, we get the system responses under the proposed impulsive control are shown in Figs. 4.5 and 4.6. From Figs. 4.5 and 4.6, we know the impulsive control protocol designed in this chapter can also realize the consensus of time-delayed MASs with stochastic perturbation in this example.

4.5 Summary

In this chapter, we have investigated the impulsive consensus problem of stochastic MASs with time delay. Based on the Lyapunov stability theory, impulsive differential equations, stochastic differential equations, and some novel delayed differential

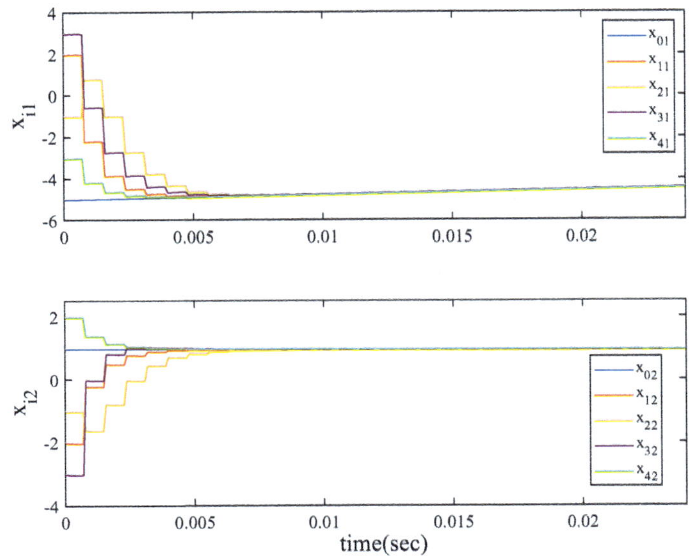

Fig. 4.6 State trajectories of $x_i(i = 0, 1, 2, 3, 4)$ for all agents in Example 4.2

inequality techniques, some sufficient conditions are obtained. Since stochastic disturbances and other channel noises are ubiquitous in both natural and manmade systems, the distributed impulsive control scheme in this chapter is more reasonable for the practical system.

References

Chen H, Shi P, Lim CC (2020) Synchronization control for neutral stochastic delay markov networks via single pinning impulsive strategy. IEEE Trans Syst Man Cybern: Syst 50(12):5406–5419

Fu X, Liu S (2023) Containment control for Riemann-Liouville fractional multi-agent singular systems with timeĺCvarying delays. Int J Dyn Control 11(4):1966–1973

Grifa MT, Pepe P (2021) On stability analysis of discrete-time systems with constrained time-delays via nonlinear Halanay-type inequality. IEEE Control Syst Lett 5(3):869–874

Hong S, Zhang Y (2022) Consensus of switched delay multi-agent systems via self-triggered impulsive control. Int J Control 95(12):3252–3261

Ke C, Li C, Zhang Q (2020) Distributed adaptive fault-tolerant consensus of nonlinear multi-agent systems via state-constraint impulsive protocols with time-delay. IEEE Trans Netw Sci Eng 7(4):3112–3121

Li Z, Wen G, Duan Z, Ren W (2015) Designing fully distributed consensus protocols for linear multi-agent systems with directed graphs. IEEE Trans Autom Control 60(4):1152–1157

Li H, Yang X, Wang S (2021) Robustness for stability and stabilization of boolean networks with stochastic function perturbations. IEEE Trans Autom Control 66(3):1231–1237

Liu K, Seuret A, Xia Y (2017) Stability analysis of systems with time-varying delays via the second-order Bessel-Legendre inequality. Automatica 76:138–142

Liu B, Sun Z, Li M, Liu DN (2022) Stabilization via event-triggered impulsive control with constraints for switched stochastic systems. IEEE Trans Cybern 52(11):11834–11846

Lyu X, Cao J, Rutkowski L, Duan P (2024) Distributed saturated impulsive control for local consensus of nonlinear time-delay multiagent systems with switching topologies. IEEE Trans Autom Control 69(2):771–782

Ma T, Fu J (2011) On the exponential synchronization of stochastic impulsive chaotic delayed neural networks. Neurocomputing 74(5):857–862

Seuret A, Liu K, Gouaisbaut F (2018) Generalized reciprocally convex combination lemmas and its application to time-delay systems. Automatica 95:488–493

Sun F, Lu C, Zhu W, Kurths J (2023) Data-sampled mean-square consensus of hybrid multi-agent systems with time-varying delay and multiplicative noises. Inf Sci 624:674–685

Tian Y, Wang Z (2021) Composite slack-matrix-based integral inequality and its application to stability analysis of time-delay systems. Appl Math Lett 120:107252

Wang X, Liu X, Zhong S, Shi L (2019) Delay-dependent impulsive distributed synchronization of stochastic complex dynamical networks with time-varying delays. IEEE Trans Syst Man Cybern: Syst 49(7):1496–1504

Wang P, Wang X, Su H (2021) Input-to-state stability of impulsive stochastic infinite dimensional systems with poisson jumps. Automatica 128:109553

Wei L, Chen WH, Luo S, Huang G (2022) Impulsive average-consensus of multi-agent systems with time-delays. J Franklin Inst 359(2):1544–1568

Xia X, Bai J, Li X, Wen G (2023) Containment control for fractional order MASs with nonlinearity and time delay via pull-based event-triggered mechanism. Appl Math Comput 454:128094

Xie X, Li X, Liu X (2023) Event-triggered impulsive control for multi-agent systems with actuation delay and continuous/periodic sampling. Chaos, Solitons Fractals 175:114067

Yang Y, Liu J, Liu B, Shi Y (2021) Mean-square exponential stability analysis and Hąxx performance of impulsive stochastic systems with time-varying uncertain parameters. Int J Innov Comput Inf Control 17(1):275–285

You L, Jiang X, Zheng S, Yan H (2024) Communication limited hybrid impulsive control of fuzzy time-delay multiagent network. IEEE Trans Fuzzy Syst 32(1):152–159

Yu L, Cui Y, Liu Y, Alotaibi ND, Alsaadi FE (2022) Sampled-based consensus of multi-agent systems with bounded distributed time-delays and dynamic quantisation effects. Int J Syst Sci 53(13):2390–2406

Zhang K, Zhang Y, Li R, Chen G (2022a) Sampled-data-based consensus of multi-agent systems with multiplicative noise and time-delays. Inf Sci 609:1621–1640

Zhang Z, Peng S, Liu D, Wang Y, Chen T (2022b) Leader-following mean-square consensus of stochastic multiagent systems with ROUs and RONs via distributed event-triggered impulsive control. IEEE Trans Cybern 52(3):1836–1849

Zhu Y, Huang Q, Wu J, Zheng Y, Wang H (2022) Synchronous and asynchronous resilient impulsive control for group consensus of second-order multi-agent systems with communication delays. ISA Trans 131:274–281

Zhuang J, Peng S, Wang Y (2022) Exponential consensus of nonlinear stochastic discrete-time multi-agent systems with time-varying delay via impulsive control. Int J Syst Sci 53(13):3286–3201

Chapter 5
Distributed Discontinuous Adaptive Impulsive Consensus of Uncertain Multi-agent Systems

5.1 Introduction

In most of the existing results of impulsive consensus of multi-agent systems(MASs), the system parameters are known. However, when designing the controller in practice, the real parameters in the system model cannot be obtained because the model is limited. For example, in engineering applications, it is impossible to obtain the system parameter values of robots, drones, grid components, and other agents in advance. Owing to its analytical complexity, further exploration of adaptive impulsive control presents great challenges and is significant both theoretically and practically. Recently, adaptive impulsive protocols have been used to achieve synchronization of chaotic systems (Xi et al. 2014; Zhang et al. 2017; Li et al. 2017) and complex networks (Wang et al. 2016; Leng and Wu 2019; Zhu et al. 2020), as well as consensus in MASs (Ma et al. 2018a, b; Xiao et al. 2016). However, in these literatures, most of the adaptive laws are continuous, which means that the state or error information of the system needs to be continuously measured during the whole control process, thus consuming a large amount of communication cost, which conflicts with the advantage of impulsive control to save communication resource consumption. Therefore, in the design of adaptive law, how to avoid the continuous communication between nodes is a key problem.

In practice, MASs are usually affected by external interference. These disturbances are usually uncertain and are therefore regarded as random factors or random effects, for example, in the current sharing control problem of DC power grids, components such as power generation units may be affected by random noise. Therefore, in the actual system, from the theoretical and practical point of view, random disturbance is a factor that cannot be ignored. To solve the above problems, in order to achieve the consensus of uncertain MASs under random disturbance and reduce the communication cost, this chapter designs an adaptive impulsive control strategy with discontinuous adaptive law and obtains sufficient conditions to achieve the con-

Z. Zhang et al., *Distributed Impulsive Coordination of Multi-Agent Systems*, Intelligent Control and Learning Systems 15, https://doi.org/10.1007/978-981-97-6262-0_5

sensus of uncertain MASs. Finally, two simulation examples are given to illustrate the effectiveness of the proposed discontinuous adaptive impulsive control strategy.

5.2 Problem Description

Consider the model of agent i as

$$dx_i(t) = (Ax_i(t) + f(x_i(t)) + u_i(t))\,dt + \vartheta_i(t, \delta_i(t))d\omega(t), \tag{5.1}$$

where $i = 1, 2, \ldots, N$, $x_i(t) \in \mathbb{R}^n$ is the state of agent i, $u_i(t) \in \mathbb{R}^n$ is the impulsive input of agent i, $A \in \mathbb{R}^{n\times n}$ is the system matrix with unknown parameter, $f(\cdot)$ is the nonlinear function which contains unknown parameter and satisfies the Lipschitz condition. $\vartheta_i(t, \delta_i(t)) : \mathbb{R}^+ \times \mathbb{R}^n \to \mathbb{R}^{n\times n}$ is the diffusion coefficient matrix, and $\omega(t)$ is the n-dimensional Brown motion.

Consider the model of leader as

$$\dot{x}_0(t) = Ax_0(t) + f(x_0(t)), \tag{5.2}$$

where $x_0(t) \in \mathbb{R}^n$ is the state of leader.

Definition 5.1 The global synchronization error for agent i is

$$\delta_i(t) = x_i(t) - x_0(t), \delta_i(t) \in \mathbb{R}^n. \tag{5.3}$$

Definition 5.2 The local neighborhood synchronization error for node i is

$$e_i(t) = \sum_{j\in N_i} a_{ij}(x_i(t) - x_j(t)) + c_i(x_i(t) - x_0(t)). \tag{5.4}$$

Assumption 5.1 Assume $\vartheta(t, 0) = 0$, and there exists a constant matrix Σ such that

$$trace[\vartheta^T(t, x(t))\,\vartheta(t, x(t))] \le \|\Sigma x(t)\|^2. \tag{5.5}$$

Definition 5.3 For the stochastic nonlinear system $dx(t) = g_1(t, x)\,dt + g_2(t, x)\,d\omega(t)$ on $t \ge t_0$ with initial value $x(t_0) \in \mathbb{R}^n$, $g_1 : \mathbb{R}^+ \times \mathbb{R}^n \to \mathbb{R}^n$, and $g_2 : \mathbb{R}^+ \times \mathbb{R}^n \to \mathbb{R}^{n\times n}$, for the nonlinear function $V(t, x)$, an operator $\mathcal{L}$ is defined as

$$\begin{aligned}\mathcal{L}V(t, x) &= V_t(t, x) + V_x(t, x)\,g_1(t, x)\\ &\quad + \frac{1}{2}trace\left[g_2^T(t, x)\,V_{xx}(t, x)\,g_2(t, x)\right],\end{aligned} \tag{5.6}$$

where $V_t(t,x) = \frac{\partial V(t,x)}{\partial t}$, $V_x(t,x) = \left(\frac{\partial V(t,x)}{\partial x_1}, \ldots, \frac{\partial V(t,x)}{\partial x_n}\right)$, $V_{xx}(t,x) = \left(\frac{\partial V(t,x)}{\partial x_i \partial x_j}\right)$. For simplicity, denote $V_t(t,x)$, $V_x(t,x)$, and $V_{xx}(t,x)$ as $V_t(t)$, $V_x(t)$, and $V_{xx}(t)$, respectively.

To achieve consensus of uncertain MASs under random disturbances, it is necessary to design appropriate distributed impulsive controllers and adaptive laws, so that under any initial conditions, the global error variable of the MASs

$$\lim_{t\to\infty} E\left(\|\delta(t)\|^2\right) = 0, \tag{5.7}$$

where $\delta(t) = [\delta_1^T(t), \delta_2^T(t), \ldots, \delta_N^T(t)]^T$.

Some lemmas related to infinite series that need to be used in this section are given as follows:

Lemma 5.1 (Rudin 1976) *Given a sequence* $\{a_n\}$, *if* $\sum_{n=1}^{\infty} a_n$ *converges, then* $\lim_{n\to\infty} a_n = 0$.

Lemma 5.2 (Rudin 1976) *For two sequences* $\{a_n\}$ *and* $\{c_n\}$, *if* $0 \le a_n \le c_n$ *for* $n \ge n_0$, *where* n_0 *is a fixed integer, and if* $\sum_{n=1}^{\infty} c_n$ *converges, then* $\sum_{n=1}^{\infty} a_n$ *converges.*

Lemma 5.3 (Rudin 1976) *Suppose* $\{a_n\}$ *is monotonic. Then* $\{a_n\}$ *converges if and only if it is bounded.*

5.3 Main Results

5.3.1 Discontinuous Adaptive Impulsive Controller Design

The distributed impulsive controller of agent i is

$$u_i(t) = \sum_{k=1}^{\infty} b_k \left(\sum_{j\in N_i} a_{ij}(x_i(t) - x_j(t)) + c_i(x_i(t) - x_0(t)) \right) \delta^*(t - t_k), \tag{5.8}$$

where $i = 1, 2, \ldots, N$, the impulsive sequence $\{t_k\}$ satisfying $0 \le t_0 < t_1 < t_2 < \cdots < t_{k-1} < t_k < \cdots$, $\lim_{k\to\infty} t_k = \infty$, $b_k \in \mathbb{R}$ is the impulsive control gain. $\delta(t)$ is the Dirac delta function, which satisfies $\delta(t) = 0$ for $t \neq 0$ and $\int_{-\infty}^{+\infty} \delta(t)\,dt = 1$. Define $\tau_k = t_k - t_{k-1}$ as the impulsive interval.

The discontinuous adaptive is designed as

$$\dot{\hat{\lambda}}(t) = \gamma \delta^T(t_{k-1}^+)\delta(t_{k-1}^+), \quad t \in (t_{k-1}, t_k], \tag{5.9}$$

where $\hat{\lambda}(t)$ is a time-varying parameter, γ is the adaptive gain, and $\hat{\lambda}(t_0) = 0$.

From (5.1) and (5.8),

$$\begin{cases} dx_i(t) = (Ax_i(t) + f(x_i(t)))dt + \vartheta_i(t, \delta_i(t))d\omega(t), \ t \neq t_k, \\ \Delta x_i(t_k) = x_i(t_k^+) - x_i(t_k^-) = b_k \left(\sum\limits_{j \in N_i} a_{ij}(x_i(t) - x_j(t)) + c_i(x_i(t) - x_0(t)) \right), \\ x_i(t_0^+) = x_i(t_0), \ t_0 \geq 0, i = 1, 2, \ldots, N, \end{cases} \tag{5.10}$$

where $\Delta x_i(t_k)$ is the jump of the state of node i at impulse instant t_k, and $x_i(t)$ is left continuous at t_k, i.e., $x_i(t_k) = x_i(t_k^-)$.

From (5.2) and (5.10), one gets the global error system of MASs as

$$\begin{cases} d\delta_i(t) = (A\delta_i(t) + f(x_i(t)) - f(x_0(t)))\, dt + \vartheta_i(t, \delta_i(t))d\omega(t), \ t \neq t_k \\ \Delta\delta_i(t_k) = b_k e_i(t_k). \end{cases} \tag{5.11}$$

From the definition of the Kronecker product, (5.11) can be rewritten as

$$\begin{cases} d\delta(t) = \left((I_N \otimes A)\delta(t) + \bar{f}(x(t), \bar{x}_0(t))\right) dt + \vartheta(t, \delta(t))d\varpi(t), \ t \neq t_k \\ \Delta\delta(t_k) = b_k((L + C) \otimes I_n)\delta(t_k), \end{cases} \tag{5.12}$$

where $\vartheta(t, \delta(t)) = \text{diag}\{\vartheta(t, \delta(t))\}$, $x(t) = [x_1^T(t), x_2^T(t), \ldots, x_N^T(t)]^T$, $\bar{x}_0(t) = 1_N \otimes x_0(t)$, $\bar{f}(x(t), \bar{x}_0(t)) = [f^T(x_1(t)) - f^T(x_0(t)), \ldots, f^T(x_N(t)) - f^T(x_0(t))]^T$, $\varpi(t) = 1_N \otimes \omega(t)$.

5.3.2 Convergence Analysis

In this section, Theorem 5.1 will provide the sufficient conditions that for adaptive parameters $\hat{\lambda}(t_k)$ and impulsive intervals τ_k.

Theorem 5.1 *Assume the MASs satisfy Assumptions 1.1, 1.2, and 5.1, if*

$$\hat{\lambda}(t_k)\tau_k + \ln(\lambda_k \xi) \leq 0, \tag{5.13}$$

where λ_k is the maximum eigenvalue of $(b_k(L + C) + I_N)^T\ (b_k(L + C) + I_N)$, and the constant $\xi > 1$, then under the distributed impulsive control law (5.8) and the discontinuous adaptive law (5.9), the consensus of MASs will be realized in the mean-squared sense.

Proof Choose the Lyapunov function as $V(t) = \delta^T(t)\delta(t)$, and the derivative of $V(t)$ along the trajectory of (5.12) is

$$dV(t) = \mathcal{L}V(t)dt + V_\delta(t)\vartheta(t, \delta(t))d\omega(t). \tag{5.14}$$

When $t \in (t_{k-1}, t_k]$, taking mathematical expectation on both sides of (5.14) and considering $E\left(V_x(t)\,\vartheta(t, x(t))\,d\omega(t)\right) = 0$, it yields

$$D^+E(V(t)) = E(\mathcal{L}V(t)), \tag{5.15}$$

where

$$\begin{aligned}\mathcal{L}V(t) = {} & V_t(t) + V_\delta(t)(A\delta(t) + \bar{f}(x(t), \bar{x}_0(t))) \\ & + \tfrac{1}{2}\text{trace}[\vartheta^T(t,\delta)V_{\delta\delta}\vartheta(t,\delta)] \\ = {} & \delta^T(t)(I_N \otimes (A + A^T))\delta(t) + 2\delta^T(t)\bar{f}(x(t), \bar{x}_0(t)) \\ & + \text{trace}[\vartheta^T(t,\delta)\vartheta(t,\delta)].\end{aligned} \tag{5.16}$$

From Assumption 5.1, one gets

$$\text{trace}[\vartheta^T(t,\delta)\vartheta(t,\delta)] \le \delta^T(t)\Sigma^T\Sigma\delta(t) \le \lambda_\Sigma V(t), \tag{5.17}$$

where λ_Σ is the maximum eigenvalue of $\Sigma^T\Sigma$.

In addition, by using the Lipschitz condition, one gets

$$\mathcal{L}V(t) = (\lambda_A + 2l + \lambda_\Sigma)V(t), \tag{5.18}$$

where λ_A is the maximum eigenvalue of $A + A^T$.

From (5.15) and (5.18),

$$D^+E(V(t)) \le (\lambda_A + 2l + \lambda_\Sigma)V(t), \tag{5.19}$$

then one has

$$E(V(t)) \le E(V(t_{k-1}^+))\exp((\lambda_A + 2l + \lambda_\Sigma)(t - t_{k-1})). \tag{5.20}$$

When $t = t_k$, from (5.12)

$$\begin{aligned} & V(t_k^+) \\ = {} & \delta^T(t_k^+)\delta(t_k^+) \\ = {} & ((b_k((L + C) \otimes I_n) + I_{nN})\delta(t_k))^T(b_k((L + C) \otimes I_n) + I_{nN})\delta(t_k) \\ = {} & \delta^T(t_k)\left(((b_k(L + C) + I_N)^T b_k((L + C) + I_N)) \otimes I_n\right)\delta(t_k) \\ \le {} & \lambda_k V(t_k).\end{aligned} \tag{5.21}$$

Take the mathematical expectation on both sides of (5.21),

$$E(V(t_k^+)) \le \lambda_k E(V(t_k)). \tag{5.22}$$

From the adaptive law (5.9),

$$\begin{aligned}\hat{\lambda}(t_1) &= \hat{\lambda}(t_0) + \int_{t_0}^{t_1} \gamma\delta^T(t_0^+)\delta(t_0^+)dt \\ &= \hat{\lambda}(t_0) + \gamma\delta^T(t_0^+)\delta(t_0^+)\tau_1 \\ &= \gamma\delta^T(t_0^+)\delta(t_0^+)\tau_1.\end{aligned} \tag{5.23}$$

When $t \in (t_1, t_2)$, $\dot{\hat{\lambda}}(t) = \gamma\delta^T(t_1^+)\delta(t_1^+)$, then

$$\begin{aligned}\hat{\lambda}(t_2) &= \hat{\lambda}(t_1) + \int_{t_1}^{t_2} \gamma\delta^T(t_1^+)\delta(t_1^+)dt \\ &= \hat{\lambda}(t_1) + \gamma\delta^T(t_1^+)\delta(t_1^+)\tau_2 \\ &= \gamma\delta^T(t_1^+)\delta(t_1^+)\tau_2.\end{aligned} \tag{5.24}$$

In general,

$$\begin{aligned}\hat{\lambda}(t_k) &= \gamma\delta^T(t_0^+)\delta(t_0^+)\tau_1 + \gamma\delta^T(t_1^+)\delta(t_1^+)\tau_2 + \cdots + \gamma\delta^T(t_{k-1}^+)\delta(t_{k-1}^+)\tau_1 \\ &= \sum_{w=1}^{k} \gamma\delta^T(t_{w-1}^+)\delta(t_{w-1}^+)\tau_w,\end{aligned} \tag{5.25}$$

which means that

$$\lim_{k\to\infty} \hat{\lambda}(t_k) = \sum_{w=1}^{\infty} \gamma\delta^T(t_{w-1}^+)\delta(t_{w-1}^+)\tau_w = \sum_{w=1}^{\infty} \gamma V(t_{w-1}^+)\tau_w. \tag{5.26}$$

Take the mathematical expectation on both sides of (5.26),

$$\lim_{k\to\infty} E(\hat{\lambda}(t_k)) = \sum_{w=1}^{\infty} \gamma\tau_w E(V(t_{w-1}^+)). \tag{5.27}$$

From Lemma 5.1, if $\lim_{k\to\infty} E(\hat{\lambda}(t_k))$ converges, then $\lim_{k\to\infty} \gamma E(V(t_{k-1}^+))\tau_k = 0$. Since $\gamma \neq 0$ and $\lim_{k\to\infty} \tau_k \neq 0$, one gets $\lim_{k\to\infty} E(V(t_{k-1}^+)) = 0$. From (5-19), when $t \in (t_{k-1}, t_k]$, $E(V(t)) \leq E(V(t_{k-1}^+))\exp((\lambda_A + 2l + \lambda_\Sigma)(t - t_{k-1}))$. Since $\exp((\lambda_A + 2l + \lambda_\Sigma)(t - t_{k-1}))$ is bounded, $\lim_{t\to\infty} E(V(t)) \leq \lim_{t\to\infty} E(V(t_{k-1}^+))\exp((\lambda_A + 2l + \lambda_\Sigma)(t - t_{k-1})) = 0$. Then the asymptotic stability in mean-squared sense of $x(t)$ is realized. Therefore, to prove the asymptotic stability of MASs in mean-squared sense, the key point is to prove $\lim_{k\to\infty} E(\hat{\lambda}(t_k))$ converges. Two cases are considered here.

Case 1: $\lim_{t\to\infty} E(\hat{\lambda}(t)) < \lambda_A + 2l + \lambda_\Sigma$, it is obvious that $\lim_{k\to\infty} E(\hat{\lambda}(t_k))$ is bounded. Because $\gamma V(t_{w-1}^+)\tau_w > 0$, $\lim_{k\to\infty} E(\hat{\lambda}(t_k))$ is monotonic. From Lemma 5.3, one gets $\lim_{k\to\infty} E(\hat{\lambda}(t_k))$ converges.

Case 2: $\lim_{t\to\infty} E(\hat{\lambda}(t)) \geq \lambda_A + 2l + \lambda_\Sigma$.

From (5.20) and (5.22),

$$\begin{aligned} E(V(t_q^+)) &\leq \lambda_q E(V(t_q)) \\ &\leq \lambda_q \exp((\lambda_A + 2l + \lambda_\Sigma)(t_q - t_{q-1}))E(V(t_{q-1}^+)) \\ &\leq E(V(t_{q-2}^+))\lambda_q \exp((\lambda_A + 2l + \lambda_\Sigma)(t_q - t_{q-1})) \\ &\quad \lambda_{q-1} \exp((\lambda_A + 2l + \lambda_\Sigma)(t_{q-1} - t_{q-2})) \\ &\leq \lambda_q \lambda_{q-1} \cdots \lambda_1 \exp((\lambda_A + 2l + \lambda_\Sigma)(t_q - t_0))E(V(t_0)) \end{aligned} \tag{5.28}$$

and

$$\begin{aligned} E(V(t_{q+1}^+)) &\leq \lambda_{q+1} E(V(t_{q+1})) \\ &= E(V(t_q^+))\lambda_{q+1} \exp((\lambda_A + 2l + \lambda_\Sigma)(t_{q+1} - t_q)). \end{aligned} \tag{5.29}$$

Since $\dot{\hat{\lambda}}(t) = \gamma V(t_{k-1}^+) \geq 0$, when $t > \tilde{t}$, $E(\hat{\lambda}(t)) \geq \lambda_A + 2l + \lambda_\Sigma$, then from (5.29)

$$E(V(t_{q+1}^+)) \leq \lambda_{q+1} E(V(t_q^+)) \exp(\hat{\lambda}(t_{q+1})(t_{q+1} - t_q)). \tag{5.30}$$

From (5.13) in Theorem 5.1, one gets

$$\lambda_{q+1} \exp(\hat{\lambda}(t_{q+1})(t_{q+1} - t_q)) \leq \frac{1}{\xi}. \tag{5.31}$$

Take the mathematical expectation on both sides of (5.31),

$$\lambda_{q+1} E\left(\exp(\hat{\lambda}(t_{q+1})(t_{q+1} - t_q))\right) \leq \frac{1}{\xi}. \tag{5.32}$$

From (5.31) and (5.32),

$$\frac{E(V(t_{q+1}^+))}{E(V(t_q^+))} \leq \lambda_{q+1} \left(\exp(\hat{\lambda}(t_{q+1})(t_{q+1} - t_q))\right) \leq \frac{1}{\xi}. \tag{5.33}$$

When $t \in (t_q, t_{q+1}]$, since $\dot{\hat{\lambda}}(t) = \gamma \delta^T(t_q^+)\delta(t_q^+) \geq 0$,

$$\begin{aligned} \hat{\lambda}(t) &= \hat{\lambda}(t_q) + \int_{t_q}^{t} \gamma \delta^T(t_q^+)\delta(t_q^+)dt \\ &\leq \hat{\lambda}(t_q) + \int_{t_q}^{t_{q+1}} \gamma \delta^T(t_q^+)\delta(t_q^+)dt \\ &= \hat{\lambda}(t_q) + \gamma \tau_{q+1} \delta^T(t_q^+)\delta(t_q^+) \\ &= \hat{\lambda}(t_q) + \gamma \tau_{q+1} V(t_q^+), \end{aligned} \tag{5.34}$$

which leads to

$$\begin{aligned}\hat{\lambda}(t_{q+1}) &= \hat{\lambda}(t_q) + \int_{t_q}^{t_{q+1}} \gamma\delta^T(t_q^+)\delta(t_q^+)dt \\ &= \hat{\lambda}(t_q) + \gamma\tau_{q+1}V(t_q^+).\end{aligned} \tag{5.35}$$

When $t \in (t_{q+1}, t_{q+2}]$,

$$\begin{aligned}\hat{\lambda}(t) &= \hat{\lambda}(t_{q+1}) + \int_{t_{q+1}}^{t} \gamma\delta^T(t_{q+1}^+)\delta(t_{q+1}^+)dt \\ &\le \hat{\lambda}(t_{q+1}) + \int_{t_{q+1}}^{t_{q+2}} \gamma\delta^T(t_{q+1}^+)\delta(t_{q+1}^+)dt \\ &= \hat{\lambda}(t_{q+1}) + \gamma\tau_{q+2}\delta^T(t_{q+1}^+)\delta(t_{q+1}^+) \\ &= \hat{\lambda}(t_{q+1}) + \gamma\tau_{q+2}V(t_{q+1}^+) \\ &= \hat{\lambda}(t_q) + \gamma\tau_{q+1}V(t_q^+) + \gamma\tau_{q+2}V(t_{q+1}^+).\end{aligned} \tag{5.36}$$

Take the mathematical expectation on both sides of (5.36),

$$\begin{aligned}E(\hat{\lambda}(t)) &= E(\hat{\lambda}(t_q) + \gamma\tau_{q+1}V(t_q^+) + \gamma\tau_{q+2}V(t_{q+1}^+)) \\ &= \hat{\lambda}(t_q) + \gamma\tau_{q+1}E(V(t_q^+)) + \gamma\tau_{q+2}E(V(t_{q+1}^+)).\end{aligned} \tag{5.37}$$

From (5.33) and (5.37),

$$E(\hat{\lambda}(t)) \le \hat{\lambda}(t_q) + \gamma\tau_{q+1}E(V(t_q^+)) + \gamma\frac{1}{\xi}\tau_{q+2}E(V(t_q^+)), \tag{5.38}$$

then one gets

$$\begin{aligned}E(\hat{\lambda}(t_{q+2})) &= E(\hat{\lambda}(t_{q+1})) + E\left(\int_{t_{q+1}}^{t_{q+2}} \gamma\delta^T(t_{q+1}^+)\delta(t_{q+1}^+)dt\right) \\ &= E(\hat{\lambda}(t_q)) + \gamma\tau_{q+1}E(V(t_q^+)) + \gamma\tau_{q+2}E(V(t_{q+1}^+)) \\ &\le E(\hat{\lambda}(t_q)) + \gamma\tau_{q+1}E(V(t_q^+)) + \gamma\frac{1}{\xi}\tau_{q+2}E(V(t_q^+)).\end{aligned} \tag{5.39}$$

In general, when $t = t_{q+m-1} (m \in \mathbb{N}^+)$,

$$E(\hat{\lambda}(t_{q+m-1})) = E(\hat{\lambda}(t_q)) + \gamma\tau_{q+1}E(V(t_q^+)) + \cdots + \gamma\tau_{q+m-1}E(V(t_{q+m-2}^+)). \tag{5.40}$$

When $t \in (t_{q+m-1}, t_{q+m}]$,

$$
\begin{aligned}
E(\hat{\lambda}(t)) &= E(\hat{\lambda}(t_{q+m-1})) + E\left(\int_{t_{q+m-1}}^{t} \gamma V(t_{q+m-1}^{+})dt\right) \\
&= E(\hat{\lambda}(t_q)) + \gamma \sum_{w=1}^{m-1} \tau_{q+w} E(V(t_{q+m-1}^{+})) + E\left(\int_{t_{q+m-1}}^{t} \gamma V(t_{q+m-1}^{+})dt\right) \\
&\leq \hat{\lambda}(t_q) + \gamma \sum_{w=1}^{m} \tau_{q+w} E(V(t_{q+m-1}^{+})) \\
&\leq \hat{\lambda}(t_q) + \gamma E(V(t_q^{+})) \sum_{w=1}^{m} \frac{\tau_{q+w}}{\xi^{w-1}},
\end{aligned}
\tag{5.41}
$$

then one has

$$
\hat{\lambda}(t_{q+m}) = \hat{\lambda}(t_q) + \gamma \sum_{w=1}^{m} \tau_{q+w} E(V(t_{q+m-1}^{+})). \tag{5.42}
$$

Denote $\tau_{\max} = \max\{\tau_{q+1}, \tau_{q+2}, \ldots, \tau_{q+m}\}$, when $t \in (t_{q+m-1}, t_{q+m}]$, from (5.41)

$$
E(\hat{\lambda}(t)) \leq E(\hat{\lambda}(t_q)) + \gamma \tau_{\max} E(V(t_q^{+})) \sum_{w=1}^{m} \frac{1}{\xi^{w-1}}, \tag{5.43}
$$

which leads to

$$
\begin{aligned}
\lim_{t\to\infty} E(\hat{\lambda}(t)) &\leq E(\hat{\lambda}(t_q)) + \gamma \tau_{\max} E(V(t_q^{+})) \sum_{w=1}^{\infty} \frac{1}{\xi^{w-1}} \\
&\leq E(\hat{\lambda}(t_q)) + \gamma \tau_{\max} E(V(t_q^{+})) \frac{\xi}{\xi - 1}.
\end{aligned}
\tag{5.44}
$$

Notice $E(V(t_q^{+}))$, $E(\hat{\lambda}(t_q))$, γ, $\tau_{\max}$, and $\xi/(\xi-1)$ are bounded. Denote

$$
\begin{aligned}
\sum\nolimits_{n=1}^{\infty} a_n &= E(\hat{\lambda}(t_q)) + \gamma \sum\nolimits_{w=1}^{\infty} \tau_{q+w} E(V(t_{q+m-1}^{+})), \\
\sum\nolimits_{n=1}^{\infty} c_n &= E(\hat{\lambda}(t_q)) + \gamma \tau_{\max} E(V(t_q^{+})) \sum\nolimits_{w=1}^{\infty} 1/\xi^{w-1}.
\end{aligned}
\tag{5.45}
$$

From (5.44), $\sum_{n=1}^{\infty} c_n$ converges to $E(\hat{\lambda}(t_q)) + \gamma \tau_{\max} E(V(t_q^{+}))\xi/(\xi-1)$. From (5.33) and the definition of $\tau_{\max}$,

$$
0 \leq \gamma \sum\nolimits_{w=1}^{\infty} \tau_{q+w} E(V(t_{q+m-1}^{+})) \leq \gamma \tau_{\max} E(V(t_q^{+})) \sum\nolimits_{w=1}^{\infty} \frac{1}{\xi^{w-1}}, \tag{5.46}
$$

then one gets $0 \leq a_n \leq c_n$. From Lemma 5.1, $\sum_{n=1}^{\infty} a_n$ converges, then $\lim_{k\to\infty} E(\hat{\lambda}(t_k))$ converges. This completes the proof of Theorem 5.1.

Table 5.1 Comparison of the adaptive laws in Xi et al. (2014), Zhang et al. (2017), Li et al. (2017), Wang et al. (2016), Leng and Wu (2019), Zhu et al. (2020), Ma et al. (2018a, b), Xiao et al. (2016), and this chapter

References	Adaptive law	Required state
Xi et al. (2014)	$\dot{\tilde{\lambda}}_l(t) = \eta e_l^2(t)$	$e_l(t)$
Zhang et al. (2017)	$\dot{d}_i(t) = k_i e_i^T(t)\Gamma e_i(t)$	$e_i(t)$
Li et al. (2017)	$\dot{\hat{L}} = \gamma\|e(t)\|^2$	$e(t)$
Wang et al. (2016)	$\dot{\hat{\xi}} = -F^T(x(t))e(t)$ $\dot{\hat{\eta}} = G(y(t))e(t)$	$e(t)$
Leng and Wu (2019)	$\dot{d}_i(t) = h_i e_i^T(t)e_i(t)$	$e_i(t)$
Zhu et al. (2020)	$\dot{\hat{L}} = \rho\sum_{k=1}^{N} E_k^T(t)E_k(t)$	$E_k(t)$
Ma et al. (2018a, b), Xiao et al. (2016)	$\dot{\hat{\lambda}}(t) = \gamma\delta^T(t)\delta(t)$	$\delta(t)$
This chapter	$\dot{\hat{\lambda}}(t) = \gamma\delta^T(t_{k-1}^+)\delta^T(t_{k-1}^+)$	$\delta(t_{k-1}^+)$

Remark 5.1 In Theorem 5.1, the choice of ξ will affect the size of the calculated impulsive interval. From (5.13), one obtains the allowable range of the impulsive interval as

$$\tau_k \leq -\frac{\ln(\xi\lambda_k)}{\hat{\lambda}(t_k)}. \tag{5.47}$$

Therefore, if ξ decreases, τ_k will increase. If $\xi\lambda_k > 1$, then from (5.37), $\tau_k < 0$, which is not feasible in practice. In order to minimize control losses while ensuring consent in uncertain MASs, one can select ξ as a consent slightly greater than 1 to obtain a larger allowable range impulsive interval. Then the impulsive interval is selected as $-\ln(\xi\lambda_k)/\hat{\lambda}(t_k)$, which is the upper bound of the allowable range.

Remark 5.2 The adaptive laws in Xi et al. (2014), Zhang et al. (2017), Li et al. (2017), Wang et al. (2016), Leng and Wu (2019), Zhu et al. (2020), Ma et al. (2018a, b), and Xiao et al. (2016) need continuously measured state or error information (see Table 5.1). Compared with these continuous adaptive laws, the discontinuous adaptive law selected in this chapter needs the measured system state or error only at impulsive instants, and the large communication costs can be reduced greatly.

Remark 5.3 From (5.21), $V(t_k^+) \leq \lambda_k V(t_k)$. Therefore, to guarantee $V(t_k)^+ \leq V(t_k)$, $\lambda_k \in (0, 1)$ is a necessary condition. From Theorem 5.1, λ_k is the maximum eigenvalue of $(b_k(L + C) + I_N)^T(b_k(L + C) + I_N)$, then one can choose the appropriate impulsive control gain b_k to satisfy $\lambda_k \in (0, 1)$.

Remark 5.4 In Gao and Hu (2015), the adaptive impulsive synchronization of uncertain systems was investigated, and its adaptive law was selected as

$$\begin{aligned} \dot{\tilde{p}}(t) &= -F^T\left(t, y\left(t_{i-1}^-\right)\right)e\left(t_{i-1}^-\right), \quad t \neq t_i^+, \\ \Delta p &= \tilde{p}\left(t_i^+\right) - \tilde{p}\left(t_i^-\right) = 0, \quad t = t_i^+, \end{aligned} \tag{5.48}$$

where the impulsive sequence $\{t_i\}$ satisfied $0 \le t_0 < t_1 < t_2 < \cdots < t_{i-1} < t_i < \cdots$, $p = (p_1, p_2, \ldots, p_n)^T \in \mathbb{R}^n$ was the unknown parameter vector. $\tilde{p} \in \mathbb{R}^n$ was the estimation of p, $e = (e_1, e_2, \ldots, e_n)^T = (y_1 - x_1, y_2 - x_2, \ldots, y_n - x_n)^T$ was the error vector, $F(\cdot) : \mathbb{R}^n \to \mathbb{R}^{n\times n}$ was a function matrix. Note that the adaptive law in Gao and Hu (2015) was discontinuous. The upper bound of impulsive interval $\omega = t_i - t_{i-1}$ was obtained as

$$0 < \omega \le \frac{1}{\Theta} \ln\left(\frac{1}{K\Phi}\right),$$

where $K > 1$, $\Theta \le 1$, $\Phi \in (0, 1)$ satisfied $V\left(t_i^+\right) = \Phi V\left(t_i^-\right)$, and $V(t)$ was the Lyapunov function. However, in Gao and Hu (2015), the value of parameter $\Phi \in (0, 1)$ was indefinite. Therefore, it is impossible to determine the upper bound of ω. Compared with Gao and Hu (2015), the parameter λ_k in this chapter is derived by the impulsive control gain b_k directly, then the upper bound of τ_k is further obtained as

$$\tau_k \le -\frac{\ln(\xi\eta_k)}{\hat{a}(t_k)}.$$

What's more, compared with Gao and Hu (2015), a new analysis frame is given in the proof of Theorem 5.1. The convergence of $\lim\limits_{k\to\infty} E(\hat{\lambda}(t_k))$ is proven firstly, then based on Lemmas 5.1–5.3, the consensus of MASs is realized. From Remark 5.1, in order to minimize the cost of impulsive control while ensuring the consensus of MASs, the impulsive interval is selected as the upper bound value within the allowable range, i.e., $\tau_k = 0.414/\hat{\lambda}(t_k)$.

Remark 5.5 In the proof process of Theorem 5.1, two possible scenarios were considered for the final value of the adaptive variable $\hat{\lambda}(t)$, namely $\lim\limits_{k\to\infty} E(\hat{\lambda}(t_k)) < \lambda_A + 2l + \lambda_\Sigma$ and $\lim\limits_{k\to\infty} E(\hat{\lambda}(t_k)) \ge \lambda_A + 2l + \lambda_\Sigma$. From the adaptive law (5.9), $\dot{\hat{\lambda}}(t) \ge 0$, so $\lim\limits_{t\to\infty} \hat{\lambda}(t)$ may be infinite or a constant value, and the magnitude of this constant value cannot be directly determined. In addition, it should be noted that (5.30) only holds when $\hat{\lambda}(t) > \lambda_A + 2l + \lambda_\Sigma$. If $\lim\limits_{k\to\infty} E(\hat{\lambda}(t_k)) < \lambda_A + 2l + \lambda_\Sigma$, then the analysis process after (5.30) will not hold. Therefore, in order to make the analysis process more rigorous, it is necessary to consider whether the consensus of MASs can be achieved in the case of $\lim\limits_{k\to\infty} E(\hat{\lambda}(t_k)) < \lambda_A + 2l + \lambda_\Sigma$.

Remark 5.6 Compared with the perturbation methods in Chap. 4, Theorem 5.1 is not necessary to assume that the constant matrix Σ in (5.30) is known. That is to say, the discontinuous adaptive impulsive control strategy designed in this chapter does not require knowledge of the values of the system matrix A and nonlinear function $f(\cdot)$, nor does it require knowledge of the parameters related to random disturbances. In addition, although the parameters λ_A and λ_Σ appear in (5.18) of the proof process of Theorem 5.1, which are the largest eigenvalues of matrices $A + A^T$ and $\Sigma^T\Sigma$, respectively, since the system matrix A and the constant matrix Σ are unknown, the parameters λ_A and λ_Σ are also unknown. It should be noted that parameters λ_A, l, and λ_Σ are only used for theoretical derivation and analysis in the proof process and are not used for designing the controller and obtaining the impulsive control interval.

Remark 5.7 The discontinuous adaptive impulsive control algorithm designed in this chapter also has its limitations and shortcomings. As shown in Eq. (5.9), the adaptive impulsive control law designed in this chapter needs to obtain the global consensus error value of the system at the impulsive instant. However, since the control strategies studied in this book are all distributed control algorithms, in the designed controller (5.8), only the neighborhood error of adjacent nodes needs to be measured at the impulsive instants. From (5.3) and (5.4), the relationship of global consensus error $\delta(t)$ and the local neighborhood consensus error $e(t)$ is

$$e(t) = ((L + C) \otimes I_n)\delta(t). \tag{5.49}$$

From Lemma 1.1 and the property (5) of the Kronecker product in Lemma 1.2, one knows the inverse of matrix $(L + C) \otimes I_n$ exists, and $((L + C) \otimes I_n)^{-1} = (L + C)^{-1} \otimes I_n$, then one gets

$$\delta(t) = ((L + C)^{-1} \otimes I_n)e(t). \tag{5.50}$$

This indicates that the global consensus error $\delta(t)$ of MASs can be calculated from the neighborhood error $e(t)$ and the Laplace matrix L and the pinning matrix C. Therefore, the discontinuous adaptive impulsive scheme designed for uncertain nonlinear MASs with random disturbances in this chapter is feasible. However, to use the discontinuous adaptive law (5.9) designed in this chapter, it is necessary to calculate the global consensus error $\delta(t)$ based on the measured neighborhood error $e(t)$ and Eq. (5.54), which adds some additional computational complexity. Moreover, since the Laplace matrix L and traction matrix C of the MASs need to be known, it also indicates that the control scheme designed in this chapter is not fully distributed. Even so, the discontinuous adaptive impulsive control strategy designed in this chapter has great advantages compared with those in literature (Ma et al. 2018a, b; Xiao et al. 2016). The adaptive impulsive control law designed in Ma et al. (2018a, b) and Xiao et al. (2016) is $\dot{\lambda}(t) = \gamma\delta^T(t)\delta(t)$, which requires continuous measurement of the global consensus error of the MASs, while the adaptive impulsive control law designed in this chapter only needs to measure the global consensus error of the MASs at some discrete control instants. Therefore, compared with the continuous adaptive law in literature (Ma et al. 2018a, b; Xiao et al. 2016), the discontinuous adaptive law designed in this chapter can greatly reduce the transmission loss of information between agents.

5.4 Simulation Results

Since this chapter considers situations where the system model contains unknown parameters and random disturbances, two different chaotic circuit system models will be used in the simulation verification in this section, namely Chua's circuit system in Example 5.1 and the Lorenz circuit system in Example 5.2, to verify that the discontinuous adaptive impulsive control algorithm designed in this chapter is applicable to different systems.

Example 5.1 Chua's circuit system.

Consider the model of each agent as

$$\begin{cases} \dot{x}_{i1} = p_1(x_{i2} - x_{i1} - g(x_{i1})), \\ \dot{x}_{i2} = x_{i1} - x_{i2} + x_{i3}, \\ \dot{x}_{i3} = -p_{i2}x_{i2}, \end{cases} \tag{5.51}$$

where p_1 and p_2 are unknown system parameters, $i = 0, 1, 2, 3, 4$, the expression of nonlinear function $g(\cdot)$ is

$$g(x_{i1}) = m_2 x_{i1} + 0.5(m_1 - m_2)\left(|x_{i1} + 1| - |x_{i1} - 1|\right), \tag{5.52}$$

where m_1 and m_2 are the unknown parameters in $g(\cdot)$.

In addition, consider the stochastic perturbation terms in the system model of agent i as $\vartheta_i(t, \delta_i) = \text{diag}\left\{1/\sqrt{2}\delta_{i1}, 1/\sqrt{2}\delta_{i2}, 1/\sqrt{2}\delta_{i3}\right\}$. Choose $\xi = 1.01$, $\gamma = 5$, $\Sigma = I_3$.

Consider the topology of MASs in Fig. 5.1. From the graph theory, one gets

$$\mathcal{A} = \begin{bmatrix} 0\,1\,0\,0 \\ 1\,0\,0\,0 \\ 0\,0\,0\,1 \\ 0\,0\,0\,0 \end{bmatrix}, \mathcal{D} = \begin{bmatrix} 1\,0\,0\,0 \\ 0\,1\,0\,0 \\ 0\,0\,1\,0 \\ 0\,0\,0\,0 \end{bmatrix}$$

$$L = \mathcal{D} - \mathcal{A} = \begin{bmatrix} 1 & -1 & 0 & 0 \\ -1 & 1 & 0 & 0 \\ 0 & 0 & 1 & -1 \\ 0 & 0 & 0 & 0 \end{bmatrix}, C = \begin{bmatrix} 1\,0\,0\,0 \\ 0\,0\,0\,0 \\ 0\,0\,0\,0 \\ 0\,0\,0\,1 \end{bmatrix}.$$

From Remark 5.3, one knows $\lambda_k \in (0, 1)$ is a necessary condition to realize consensus of MASs. From the definition of λ_k, one gets λ_k is related to impulsive control gain b_k, the relationship of λ_k and b_k is shown in Fig. 5.2. From Fig. 5.2, one knows the selection range of b_k is $(-0.764, 0)$. Moreover, when $b_k = -0.691$, the value of λ_k is minimum. From (5.13), when other parameters remain unchanged, the smaller the value of λ_k, the obtained impulsive interval is larger, and the required control loss is smaller. Therefore, the impulsive control gain is chosen as $b_k = -0.691$.

In this example, select the initial values of each agent in MASs as $x_0 = [2, 0, -1]^T$, $x_1 = [-3, 1, 2]^T$, $x_2 = [-2, -1, 2]^T$, $x_3 = [-1, -1, 1]^T$, $x_4 = [1, 1, 1]^T$. From

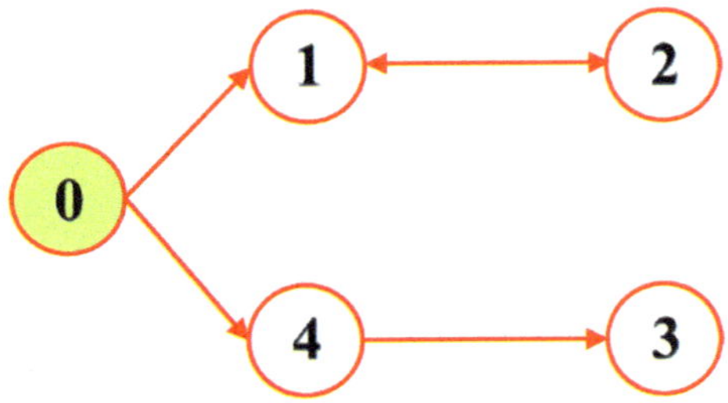

Fig. 5.1 Topology graph of the multi-agent systems in Example 5.1

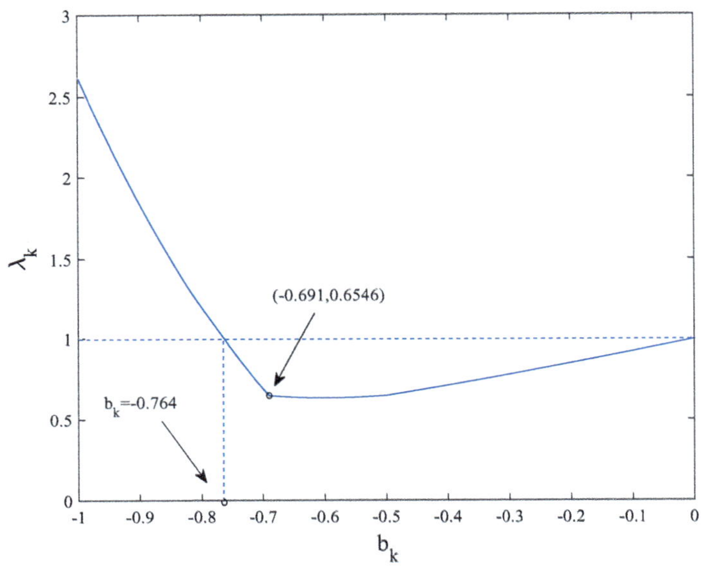

Fig. 5.2 Allowable range of impulsive control gain b_k in Example 5.1

(5.13) in Theorem 5.1, one obtains the allowable range of impulsive control interval τ_k is

$$\tau_k \leq -\frac{\ln(\lambda_k \xi)}{\hat{\lambda}(t_k)} = \frac{0.414}{\hat{\lambda}(t_k)}. \tag{5.53}$$

Based on the selected system and control parameters, the curve of the adaptive parameter $\hat{\lambda}(t)$ is shown in Fig. 5.3. From Fig. 5.3, it can be seen that the parameter $\hat{\lambda}(t)$ initially increases with the increase of t, because $\dot{\hat{\lambda}}(t) = \gamma\delta^T(t_{k-1}^+)\delta(t_{k-1}^+) \geq 0$. Note that the parameter $\hat{\lambda}(t)$ in Fig. 5.3 does not become infinite when $t \to \infty$, but tends to a constant, which is consistent with the theoretical analysis results in this chapter. The relationship between τ_k and k is presented in Fig. 5.4. From Fig. 5.4,

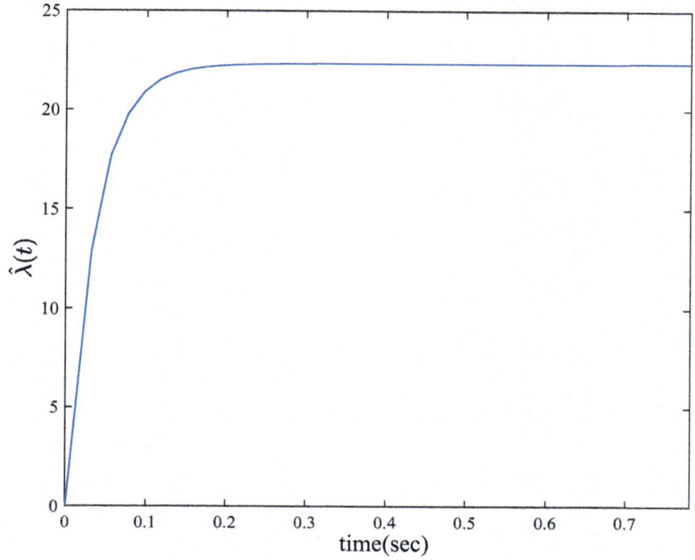

Fig. 5.3 The evolution of adaptive parameter $\hat{\lambda}(t)$ in Example 5.1

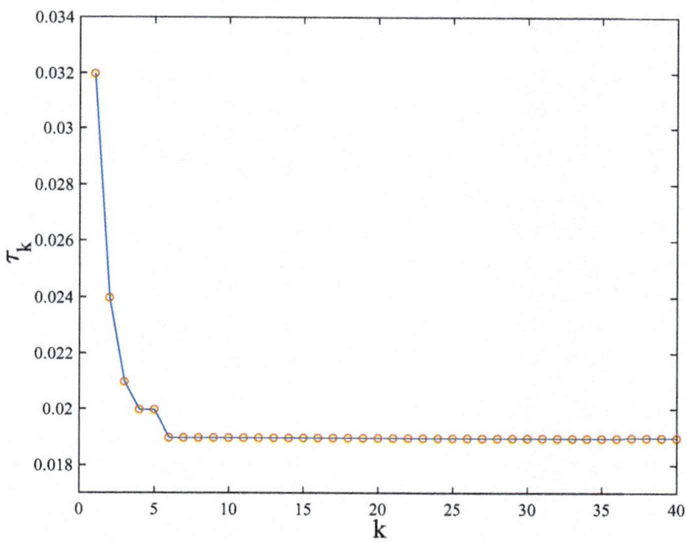

Fig. 5.4 The impulsive interval τ_k versus k in Example 5.1

one gets the τ_k is monotonically nonincreasing because $\hat{\lambda}(t_k)$ is monotonically nondecreasing. Furthermore, since $\hat{\lambda}(t_k)$ converges to a positive constant when $k \to \infty$, the impulsive interval τ_k does not decrease to 0 but converges to a positive constant

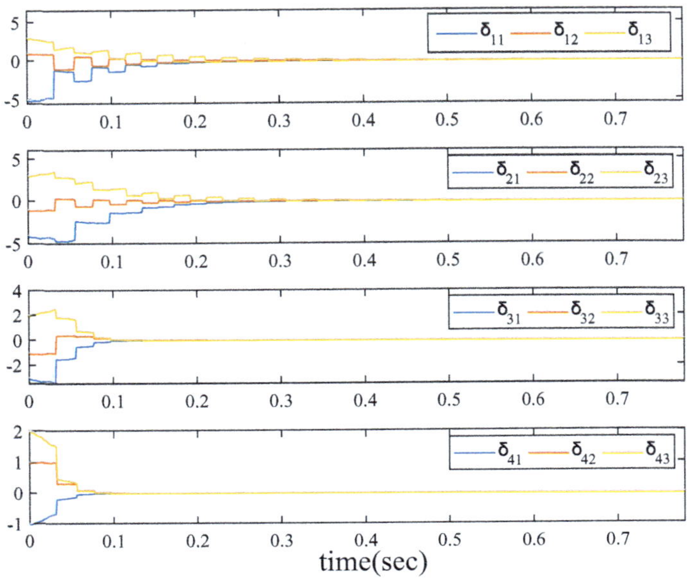

Fig. 5.5 Global consensus error of uncertain nonlinear multi-agent systems in Example 5.1

when $k \to \infty$. Figure 5.5 shows the curve of global consensus error $\delta(t)$. As shown in Fig. 5.5, under the discontinuous adaptive impulsive control law designed in this chapter, the state of each follower agent can achieve consensus with the corresponding state of the leader, that is, the consensus of MASs in the mean-squared sense can be achieved, which demonstrates the effectiveness of Theorem 5.1.

It should be noted that the objective of the discontinuous adaptive impulsive control strategy designed in this chapter is only to ensure the consensus of the MASs, but it cannot guarantee the estimation of the true parameter values of the system. How to accurately estimate the system parameters of nonlinear uncertain MASs is also a challenging problem that needs to be studied in the future.

Example 5.2 Lorenz circuit system.
In this example, the model of agent is considered as the Lorenz circuit system (Liu and Chen 2008) as follows:

$$\begin{cases} \dot{x}_{i1} = \sigma(x_{i2} - x_{i1}) \\ \dot{x}_{i2} = s x_{i1} - x_{i1} x_{i3} - x_{i2} \\ \dot{x}_{i3} = x_{i1} x_{i2} - \beta x_{i3}, \end{cases} \tag{5.54}$$

where σ, s, and β are unknown parameters, and the random disturbance term present in the follower agent is the same as in Example 5.1. The topology diagram of the MASs in this example is shown in Fig. 5.6, which is different from the topology

Fig. 5.6 Topology graph of the multi-agent systems in Example 5.2

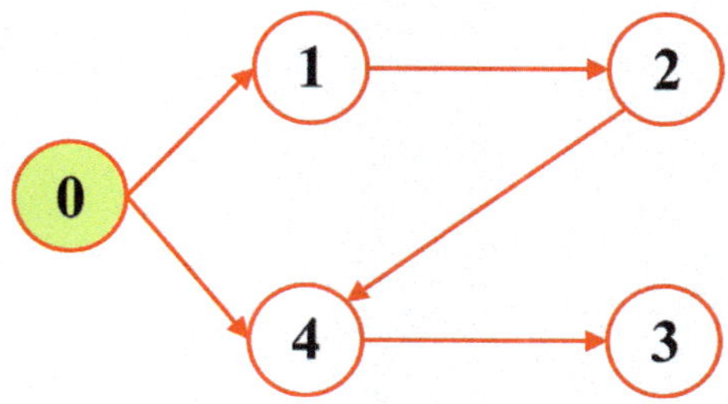

considered in Example 5.1, but still satisfies the directed spanning tree condition. From the graph theory, one can get

$$\mathcal{A} = \begin{bmatrix} 0\,0\,0\,0 \\ 1\,0\,0\,0 \\ 0\,0\,0\,1 \\ 0\,1\,0\,0 \end{bmatrix}, \mathcal{D} = \begin{bmatrix} 0\,0\,0\,0 \\ 0\,1\,0\,0 \\ 0\,0\,1\,0 \\ 0\,0\,0\,1 \end{bmatrix},$$

$$L = \mathcal{D} - \mathcal{A} = \begin{bmatrix} 0 & 0 & 0 & 0 \\ -1 & 1 & 0 & 0 \\ 0 & 0 & 1 & -1 \\ 0 & -1 & 0 & 1 \end{bmatrix}, C = \begin{bmatrix} 1\,0\,0\,0 \\ 0\,0\,0\,0 \\ 0\,0\,0\,0 \\ 0\,0\,0\,1 \end{bmatrix}.$$

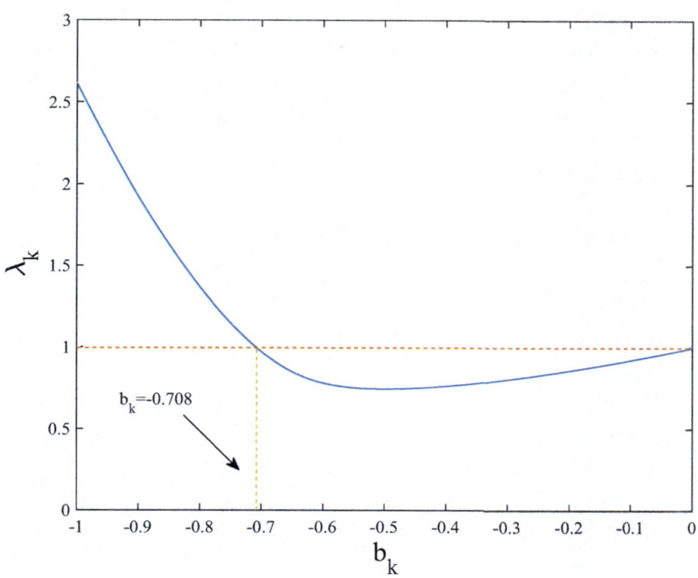

Fig. 5.7 Allowable range of impulsive control gain b_k in Example 5.2

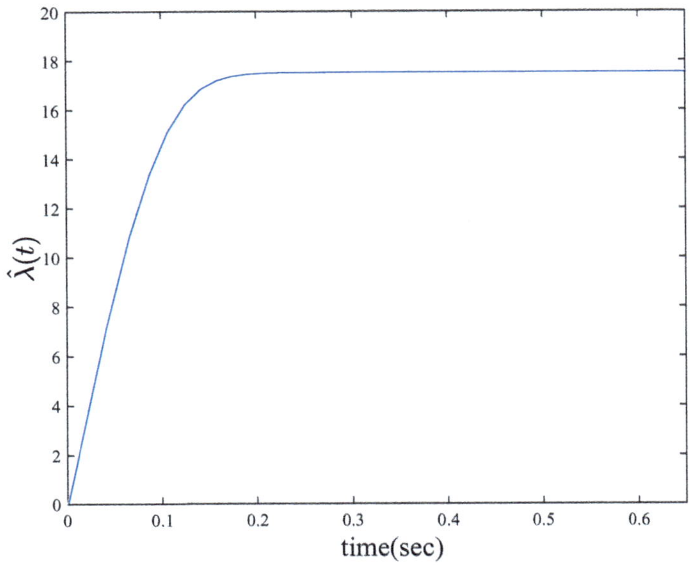

Fig. 5.8 The evolution of adaptive parameter $\hat{\lambda}(t)$ in Example 5.2

In this example, the relationship between λ_k and b_k is presented in Fig. 5.7. From Fig. 5.7, one knows when $b_k \in (-0.708, 0)$, $\lambda_k \in (0, 1)$. From Remark 5.3, the selectable range of impulsive control gain is $b_k \in (-0.708, 0)$. In addition, when $b_k = 0.5$, λ_k is smallest. Similar to Example 5.1, to obtain a larger impulsive interval for saving control lost, we choose $b_k = 0.5$ in this example. Moreover, we select $\xi = 1.01$, and the initial values of each agent in MASs are $x_0 = [2, 0, -1]^T$, $x_1 = [3, -1, 1]^T$, $x_2 = [-1, -2, 1]^T$, $x_3 = [1, -1, 1]^T$, $x_4 = [1, 1, 1]^T$.

The evolution of $\hat{\lambda}(t)$ is presented in Fig. 5.8, which shows that $\lim_{k\to\infty} E(\hat{\lambda}(t_k))$ is monotonically nondecreasing and converges to a positive constant. The curve of the impulsive interval τ_k is shown in Fig. 5.9. From Fig. 5.9, one gets τ_k converges to a positive constant when $k \to \infty$. Figure 5.10 is the curve of the global consensus error of the MASs over time in Example 5.2. It can be seen that the adaptive impulsive control algorithm designed in this chapter can also achieve the control objective by applying it to the Lorenz system.

Comparing Figs. 5.8 and 5.9 with Figs. 5.3 and 5.4, it can be observed that due to changes in the system model and topology, the values of adaptive parameter $\hat{\lambda}(t)$ and impulsive interval τ_k may be different, but the trends of curves are the same, which is consistent with the theoretical analysis results of this chapter. Comparing Fig. 5.10 with Fig. 5.5, it can be seen that although there are significant differences in the trajectories of the consensus error under different system models and topologies, the consensus of MASs can be achieved under the discontinuous adaptive impulsive control strategy designed in this chapter.

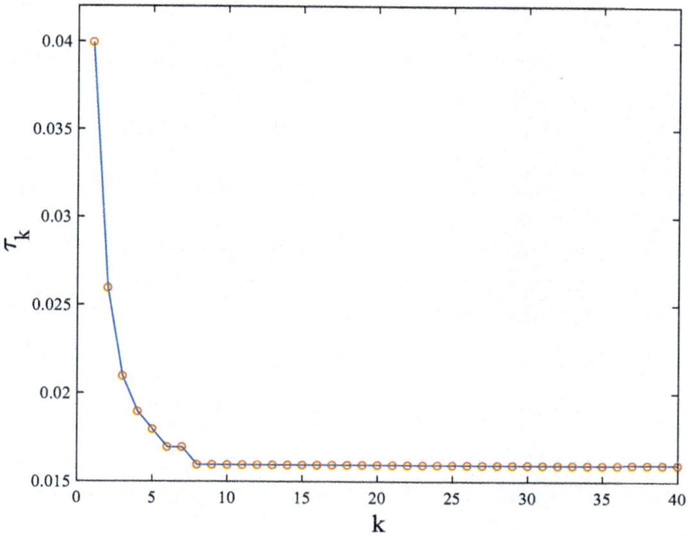

Fig. 5.9 The impulsive interval τ_k versus k in Example 5.2

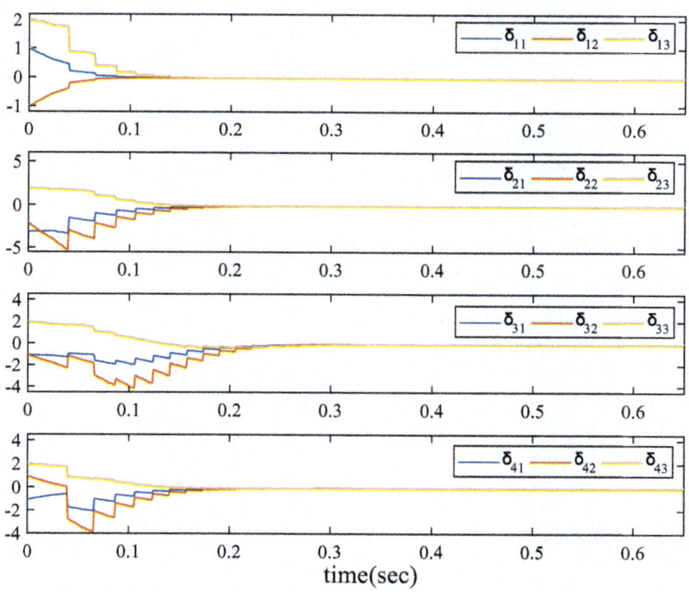

Fig. 5.10 Global consensus error of uncertain nonlinear multi-agent systems in Example 5.2

5.5 Summary

This chapter proposes a discontinuous adaptive impulsive control scheme for uncertain MASs under stochastic disturbances, targeting the limited system model, i.e., unknown system parameters and random disturbances. The results indicate that using discontinuous adaptive law can avoid continuous communication, and compared with continuous adaptive law, discontinuous adaptive law requires less communication cost. Based on the designed discontinuous adaptive pulse strategy, the consensus of MASs in the mean-squared sense is achieved. Finally, the correctness and practicality of our results were verified through two numerical examples on different systems.

References

Gao X, Hu H (2015) Adaptive-impulsive synchronization and parameters estimation of chaotic systems with unknown parameters by using discontinuous drive signals. Appl Math Model 39(14):3980–3989

Leng H, Wu Z (2019) Impulsive synchronization of complex-variable network with distributed time delays. Physica A 536:122602

Li XF, Chu YD, Leung AYT, Zhang H (2017) Synchronization of uncertain chaotic systems via complete-adaptive-impulsive controls. Chaos, Solitons Fractals 100:24–20

Liu X, Chen T (2008) Boundedness and synchronization of y-coupled Lorenz systems with or without controllers. Physica D 237(5):630–639

Ma T, Li T, Cui B (2018a) Adaptive-impulsive consensus of multiagent systems. Int J Robust Nonlinear Control 28(6):2031–2046

Ma T, Yu T, Cui B (2018b) Adaptive synchronization of multi-agent systems via variable impulsive control. J Franklin Inst 355(15):7490–7508

Rudin M (1976) Principles of mathematical analysis, 3rd edn. McGraw Hill

Wang X, She K, Zhong S, Cheng J (2016) Synchronization of complex networks with non-delayed and delayed couplings via adaptive feedback and impulsive pinning control. Nonlinear Dyn 86(1):165–176

Xi H, Yu S, Zhang R, Xu L (2014) Adaptive impulsive synchronization for a class of fractional-order chaotic and hyperchaotic systems. Optik 125(9):2036–2040

Xiao P, Ma T, Xue F, Gu Z, Fu J (2016) Adaptive impulsive consensus of multi-agent systems with unknown parameters. In: Proceedings of the 28th Chinese control and decision conference (CCDC), pp 1741–1746

Zhang X, Li D, Zhang X (2017) Adaptive fuzzy impulsive synchronization of chaotic systems with random parameters. Chaos, Solitons Fractals 104:77–83

Zhu D, Wang R, Liu C, Duan J (2020) Synchronization of chaotic-oscillation permanent magnet synchronous generators networks via adaptive impulsive control. IEEE Trans Circuits Syst II Express Briefs 67(10):2194–2198

Chapter 6
Distributed Impulsive Consensus of Fractional-Order Multi-agent Systems with Control Gain Error

6.1 Introduction

Many dynamical characteristics of natural phenomenon are more precise to be described by fractional-order dynamic equations, which have been studied for more than 300 years. The corresponding examples include chemotaxis behavior, food seeking of microbes, and the collective motion of bacteria in lubrications perspired by themselves, where the behaviors of the microbes and bacteria are described by fractional-order dynamics (Cohen et al. 2001; Kozlovsky et al. 1999). The main advantage of fractional-order systems is that the fractional-order differential operator is nonlocal and provides an excellent instrument for the description of memory and hereditary properties of various materials and processes. Therefore, it is very important and significant to study the coordination of fractional-order MASs.

In recent years, distributed collaborative control of fractional-order nonlinear systems is wildly used in biological, physical, and engineering systems, such as cluster problems and the formation of robotic systems. The consensus of fractional-order MASs was first proposed and studied in Cao et al. (2008, 2010), where the distributed coordination of networked fractional-order systems over a directed interaction graph and the distributed formation control of fractional-order systems under dynamic interaction and absolute/relative damping are studied, respectively. After this, many control techniques and significant results have been developed for the coordination of linear and nonlinear MASs. For example, Sun et al. (2024) studied group consensus for heterogeneous delayed fractional-order MASs under the cooperation-competition networks. Yan et al. (2024) investigated the consensus of fractional-order MASs modeling partial differential equations–ordinary differential equations by boundary protocol. Qing et al. (2024) researched the practical time-boundary consensus for fractional-order MASs with well-known and estimable topology.

Compared with these existing continuous control methods, impulsive control is an efficient method to deal with the dynamical systems which cannot be controlled

Z. Zhang et al., *Distributed Impulsive Coordination of Multi-Agent Systems*, Intelligent Control and Learning Systems 15, https://doi.org/10.1007/978-981-97-6262-0_6

by continuous control methods. In the literatures dealing with the impulsive consensus of fractional-order MASs, several topics have been addressed. Xiao et al. (2017a) studied the consensus of nonlinear fractional-order MASs via distributed impulsive control method. Lin et al. (2022) investigated the consensus of MASs via event-triggered impulsive control and pinning control. Yang et al. (2023) studied the impulsive consensus problem of linear MASs composed of fractional-order conformable differential equations. Narayanan et al. (2022) studied the problem of fractional-order impulsive security control for uncertain fractional-order delayed MASs under Denial-of-Service attack. Obviously, in coordinated problems of MASs, using fractional-order dynamics to describe research objects with fractional-order characteristics will better reveal the essence and dynamic behavior of the objects, so that the designed distributed coordinated controller can play a role accurately and reliably.

For the convenience of applications, impulsive control gain is often chosen as a constant value at certain impulsive instants. However, the disturbance often occurs at the impulsive control instants, which has great effect on the synchronization performance. There tends to exist impulsive control gain error caused by mechanical limitation and device restriction inevitably when impulsive control is applied in a practical system. It is difficult to alter state variables to certain values precisely at the impulsive moment (Ma and Zhao 2014). Thus, it is more important to consider control gain error with an acceptable boundary in an impulsive control system, then the results can be more significant from theoretical and practical points of view. There are a few works about the consensus of MASs under impulsive control gain error. Ma et al. (2016a, b) firstly studied the consensus problem of nonlinear MASs via impulsive control with control gain error. Zhang et al. (2017) investigated the adaptive impulsive control of MASs with control gain error. Xiao et al. (2017b) studied the impulsive consensus of MASs with parameter uncertainty and control gain error.

Based on the preceding discussions, the impulsive consensus of fractional-order MASs with control gain error is investigated in this chapter. Considering the impact of control gain error, based on the theory of impulsive differential equations, the Lyapunov stability theory, and the Mittag–Leffler function, some novel sufficient conditions for achieving the consensus of a class of fractional-order nonlinear MASs are derived. Finally, two numerical simulations are verified to illustrate the effectiveness and feasibility of the proposed method.

6.2 Problem Description

Definition 6.1 The Caputo fractional-order derivative of order $\alpha > 0$ for a function $f(t) \in ([t_0, +\infty), \mathbb{R})$ is defined as

$$
{}_{t_0}D_t^{\alpha} f(t) = \frac{1}{\Gamma(n-\alpha)} \int_{t_0}^{t} \frac{f^{(n)}(\tau)}{(t-\tau)^{\alpha+1-n}} d\tau, \tag{6.1}
$$

where n denotes a positive integer and α represents a real number that $n-1<\alpha<n$, and $\Gamma(\cdot)$ represents the Gamma function $\Gamma(p)=\int_0^{+\infty} t^{p-1}e^{-t}dt$.

Lemma 6.1 (Podlubny 1999) *Let $V(t)$ be a continuous function on $[t_0,+\infty)$ and satisfies*

$$D_t^\alpha V(t) \leq \eta V(t), \tag{6.2}$$

where $0<\alpha<1$ and $\eta \in \mathbb{R}$.

Denote $E_\alpha(z)=\sum\limits_{k=0}^{+\infty}\frac{z^k}{\Gamma(\alpha k+1)}$ is the Mittag–Leffler function, then

$$V(t) \leq V(t_0)E_\alpha(\eta(t-t_0)^\alpha). \tag{6.3}$$

Lemma 6.2 (Yu et al. 2015) *For $0<\alpha<1$ and $x \in \mathbb{R}$, $x>0$, $E_\alpha(x)$ is a monotone increasing function.*

Property 6.1 For any constants η and β, it holds that

$${}_{t_0}D_t^\alpha(\eta f(t)+\beta g(t)) = \eta\,{}_{t_0}D_t^\alpha f(t) + \beta\,{}_{t_0}D_t^\alpha g(t). \tag{6.4}$$

Lemma 6.3 *Let $x(t) \in \mathbb{R}$ be a continuous and derivable function. Then, for any time instant $t \geq t_0$*

$$\frac{1}{2}\,{}_{t_0}D_t^\alpha x^2(t) \leq x(t)\,{}_{t_0}D_t^\alpha x(t). \tag{6.5}$$

Consider the dynamics of N follower agents as

$$D^\alpha x_i(t) = Ax_i(t) + f(x_i(t)) + u_i(t), \tag{6.6}$$

where $x_i(t)=[x_{i1}(t),\ldots,x_{in}(t)]^T \in \mathbb{R}^n$ is the state of node i, $A \in \mathbb{R}^{n\times n}$, $f:\mathbb{R}^n \to \mathbb{R}^n$ is a continuous nonlinear function of agent i, $u_i(t) \in \mathbb{R}^n$ is the control input of agent i.

Consider the dynamics of the leader as

$$D^\alpha x_0(t) = Ax_0(t) + f(x_0(t)), \tag{6.7}$$

where $x_0(t) \in \mathbb{R}^n$ is the state of the leader.

Definition 6.2 The global synchronization error is

$$\delta_i(t) = x_i(t) - x_0(t),\ \delta_i(t) \in \mathbb{R}^n. \tag{6.8}$$

Definition 6.3 The local neighborhood synchronization error for agent i is

$$e_i(t) = \sum_{j\in N_i} a_{ij}(x_i(t)-x_j(t)) + c_i(x_i(t)-x_0(t)). \tag{6.9}$$

The control objective is to design distributed impulsive controller for all followers, such that all agents synchronize to the state of the leader for any initial conditions

$$\lim_{t\to\infty} \delta(t) = 0. \tag{6.10}$$

For simplicity, in the rest of this chapter, all variables will be presented without the t, i.e., $x := x(t)$, $x_i := x_i(t)$, $\delta := \delta(t)$, $\delta_i := \delta_i(t)$.

6.3 Main Results

6.3.1 Impulsive Controller Design

The impulsive control input of agent i is designed as

$$u_i(t) = \sum_{k=1}^{\infty}(b_k + \Delta b_k)\left(\sum_{j\in N} a_{ij}e_i(t)\right)\delta^*(t - t_k), \tag{6.11}$$

where b_k is the impulsive control gain, $\Delta b_k = m\varphi(t_k)b_k$ is the control gain error, $m \in (0, 1)$ is a known constant, $|\varphi(t_k)| < 1$. The Dirac delta function $\delta^*(t)$ satisfies $\delta^*(t) = 0$ if $t \neq 0$, and $\int_{-\infty}^{+\infty} \delta^*(t)dt = 1$. In this chapter, $x_i(t)$ satisfies left continuous at t_k, i.e., $x_i(t_k) = x_i(t_k^-)$.

From (6.6) and (6.11), the distributed impulsive controlled system is denoted as

$$\begin{cases} D^\alpha x_i(t) = Ax_i(t) + f(x_i(t)), \quad t \neq t_k, \\ \Delta x_i(t_k) = x_i(t_k^+) - x_i(t_k^-) = (b_k + \Delta b_k)e_i(t_k) \\ x_i(t_0^+) = x_i(t_0), \quad t_0 \geq 0, \end{cases} \tag{6.12}$$

where $\Delta x_i(t_k)$ is the jump of the state of node i at impulse instant t_k.

From (6.6), (6.7), and (6.12) the error system is obtained as

$$\begin{cases} D^\alpha \delta_i(t) = A\delta_i(t) + f(x_i(t)) - f(x_0(t)), t \neq t_k, \\ \Delta\delta_i(t_k) = (b_k + \Delta b_k)e_i(t_k). \end{cases} \tag{6.13}$$

By using the Kronecker product, the system (6.13) can be rewritten as

$$\begin{cases} D^\alpha \delta(t) = (I_N \otimes A)\delta(t) + \bar{f}(x(t), \bar{x}_0(t)), \quad t \neq t_k, \\ \Delta\delta(t_k) = (b_k + \Delta b_k)((L + C) \otimes I_n)\delta(t_k), \end{cases} \tag{6.14}$$

where $x(t) = [x_1^T(t), x_2^T(t), \ldots, x_n^T(t)]^T \in \mathbb{R}^{Nn}$, $\bar{x}_0(t) = 1_N \otimes x_0(t) \in \mathbb{R}^{Nn}$, $\bar{f}(x(t), \bar{x}_0(t)) = [f^T(x_1(t)) - f^T(x_0(t)), \ldots, f^T(x_N(t)) - f^T(x_0(t))]^T \in \mathbb{R}^{Nn}$.

6.3.2 Convergence Analysis

In this section, based on the Lyapunov theory, the main results will be given to realize the consensus of fractional-order nonlinear MASs via impulsive protocol with control gain error.

Theorem 6.1 *Assume that Assumptions 1.1 and 1.2 hold, if there exist* $\xi > 1$ *that*

$$E_\alpha((\lambda_A + 2l)(t_k - t_{k-1})^\alpha)\lambda_k \xi \leq 1, \tag{6.15}$$

where the parameters λ_A *and* λ_k *are the maximum eigenvalue of* $A + A^T$ *and* $((b_k + \Delta b_k)H + I_N)^T((b_k + \Delta b_k)H + I_N)$, *respectively,* $H = L + C$, *then the consensus of fractional-order MASs can be realized.*

Proof Choose the Lyapunov functions as $V(t) = \delta^T \delta$, for $t \in (t_{k-1}, t_k]$,

$$\begin{aligned} D^\alpha V(t) &= D^\alpha(\delta^T \delta) \\ &= 2\delta^T D^\alpha \delta \\ &= 2\delta^T((I_N \otimes (A + A^T))\delta + 2\delta^T \bar{f}(x, \bar{x}_0) \\ &\leq (\lambda_A + 2l)\delta^T \delta \\ &= (\lambda_A + 2l)V(t). \end{aligned} \tag{6.16}$$

From Lemma 6.1, for $t \in (t_{k-1}, t_k]$, we can get

$$V(t_k) \leq V(t_{k-1}^+)E_\alpha(\lambda(t_k - t_{k-1})^\alpha). \tag{6.17}$$

When $t = t_k$, from (6.14), we have

$$\begin{aligned} &V(t_k^+) \\ &= \delta^T(t_k^+)\delta(t_k^+) \\ &= (((b_k + \Delta b_k)(H \otimes I_n) + I_{nN})\delta(t_k))^T((b_k + \Delta b_k)(H \otimes I_n) + I_{nN})\delta(t_k) \\ &= \delta^T(t_k)\big((((b_k + \Delta b_k)H + I_N)^T(b_k + \Delta b_k)(H + I_N)\big) \otimes I_n\big)\delta(t_k) \\ &\leq \lambda_k V(t_k). \end{aligned} \tag{6.18}$$

When $t \in (t_0, t_1]$, from (6.17), one gets

$$V(t) \leq V(t_0)E_\alpha((\lambda_A + 2l)(t - t_0)^\alpha), \tag{6.19}$$

which leads that

$$V(t_1) \leq V(t_0)E_\alpha((\lambda_A + 2l)(t_1 - t_0)^\alpha). \tag{6.20}$$

When $t \in (t_1, t_2]$, from (6.17), (6.18), and (6.20), one gets

$$\begin{aligned} V(t) &\le V(t_1^+)E_\alpha((\lambda_A + 2l)(t - t_1)^\alpha) \\ &= \lambda_1 E_\alpha((\lambda_A + 2l)(t_1 - t_0)^\alpha) E_\alpha((\lambda_A + 2l)(t - t_1)^\alpha) V(t_0). \end{aligned} \tag{6.21}$$

In general, for $t \in (t_k, t_{k+1}]$,

$$\begin{aligned} V(t) \le\ & \lambda_1 E_\alpha((\lambda_A + 2l)(t_1 - t_0)^\alpha) \cdots \lambda_k E_\alpha((\lambda_A + 2l)(t_k - t_{k-1})^\alpha) \\ & E_\alpha((\lambda_A + 2l)(t - t_k)^\alpha). \end{aligned} \tag{6.22}$$

From (6.14) in Theorem 6.1,

$$V(t) \le \frac{1}{\xi^k} E_\alpha((\lambda_A + 2l)(t - t_k)^\alpha) V(t_0). \tag{6.23}$$

Since $(\lambda_A + 2l)$, $t - t_k$ and $V(t_0)$ are finite constants, and $\frac{1}{\xi^k} \to 0$ as $k \to \infty$ (i.e., $t \to \infty$). It is easy to conclude that δ globally converges to zero, namely, all follower x_i globally synchronize to x_0. This completes the proof.

Remark 6.1 The condition (6.14) in Theorem 6.1 can be transformed into the following form:

$$\frac{1}{\lambda_k E_\alpha((\lambda_A + 2l)(t_k - t_{k-1})^\alpha)} \ge \xi. \tag{6.24}$$

That is to say, the consensus condition is that the constraint combination

$$\frac{1}{\lambda_k E_\alpha((\lambda_A + 2l)(t_k - t_{k-1})^\alpha)},$$

which includes system parameters $(\lambda_A + 2l)$, impulsive control gain b_k, control gain error Δb_k, impulsive interval $t_k - t_{k-1}$, and the system topology connection $L + C$, is greater than a finite constant $\xi > 1$. It follows that the constant ξ contributes to the recursive result (6.22) and the convergence of synchronization error, which shows the existence and necessity of ξ in the consensus condition.

If the distributed impulsive protocol is without control gain error, the following Corollary 6.1. will be derived.

Corollary 6.1 *Assume that Assumptions 1.1 and 1.2 hold, if there exist $\xi > 1$ that*

$$E_\alpha((\lambda_A + 2l)(t_k - t_{k-1})^\alpha)\lambda_k \xi \le 1, \tag{6.25}$$

where λ_k is the maximum eigenvalue of $(b_k(L + C) + I_N)^T(b_k(L + C) + I_N)$, λ_A is same as Theorem 6.1. Then the consensus of MASs can be realized.

Conducting $m = 0$, the proof is similar to the proof of Theorem 6.1, which is omitted here. This completes the results.

Remark 6.2 For Theorem 6.1, there exists one important problem to be considered. Note that the condition in Theorem 6.1 needs the value of $\varphi(t_k)$ at every impulsive instant. However, as the stochastic characterization factor of impulsive control gain error, the value of $\varphi(t_k)$ is hard to know in advance. Therefore, how to get one criterion without the information $\varphi(t_k)$ to realize the consensus is more important and significant. From (6.18), $V(t_k^+) \leq \lambda_k V(t_k)$. Therefore, to guarantee $V(t_k^+ \leq V(t_k)$, $\lambda_k \in (0, 1)$ is a necessary condition. From Theorem 6.1, one gets the parameter λ_k is the maximum eigenvalue of $\left(b_k^*(L + C) + I_N\right)^T \left(b_k^*(L + C) + I_N\right)$, where $b_k^* = b_k + \Delta b_k = (1 + m\varphi(t_k))b_k \in (\rho_1, \rho_2)$. Therefore, we need to obtain the allowable range of b_k^* (Correspond to obtain the value of ρ_1 and ρ_2), such that $\lambda_k \in (0, 1)$ holds. Then we can obtain the reasonable b_k and m to realize the consensus.

In the following, to overcome the restriction of $\varphi(t_k)$, we give another new theorem.

Theorem 6.2 *Assume that Assumptions 1.1 and 1.2 hold. If there exists* $\xi > 1$ *such that*

$$E_\alpha((\lambda_A + 2l)(t_k - t_{k-1})^\alpha)\lambda_M \xi \leq 1, \tag{6.26}$$

$$\frac{\rho_1}{1+m} < b_k < \frac{\rho_2}{1-m}, \tag{6.27}$$

$$m < \frac{\rho_1 - \rho_2}{\rho_1 + \rho_2}, \tag{6.28}$$

where $\lambda_M = \max_{k\in\mathbb{N}_+}\{\lambda_k\}$, λ_A *and* λ_k *have the same definition in Theorem 6.1,* ρ_1 *and* ρ_2 *are the lower and the upper bounds of* b_k^* *under the condition* $\lambda_k \in (0, 1)$. *Then the consensus of fractional-order MASs is realized via distributed impulsive control with control gain error.*

Proof To realize the consensus of MASs, the condition (6.26) is necessary, which implies the further sufficient condition $\lambda_k \in (0, 1)$. It is not hard to get the lower and the upper bounds ρ_1 and ρ_2 of b_k^*. If the control gain with error $b_k^* = b_k + \Delta b_k$ is always in the range (ρ_1, ρ_2), the impulsive consensus of MASs will be realized. Since $m \in (0, 1)$ and $|\varphi(t_k)| < 1$, one gets $1 - m < \varphi(t_k) < 1 + m$. Obviously, the condition (6.27) can guarantee $b_k^* = (1 + m\varphi(t_k))b_k \in (\rho_1, \rho_2)$. The condition (6.28) is given to ensure that the inequality (6.26) must have a solution.

The rest main proof can refer to the ones in Theorem 6.1, which is omitted here. This completes the proof.

Remark 6.3 The following steps are given to use Theorem 6.2:
Step 1: Obtain the allowable range of b_k^* (Correspond to obtain the value of ρ_1 and ρ_2) based on $\lambda_k \in (0, 1)$.
Step 2: From (6.27) and (6.28), select the appropriate b_k and m.
Step 3: According to the selected parameters b_k and m, the range of $b_k^* = b_k + \Delta b_k = (1 + m\varphi(t_k))b_k$ can be calculated, and then the value of λ_M is obtained according to the relationship between b_k^* and λ_k.
Step 4: Based on λ_M obtained in Step 3, the allowable range of impulsive interval $t_k - t_{k-1}$ can be calculated.

6.4 Simulation Results

The chaotic commensurate fractional-order Chua's system is described as follows:

$$\begin{cases} D^{\alpha}x_{i1} = p_1(x_{i2} - x_{i1} - g(x_{i1})), \\ D^{\alpha}x_{i2} = x_{i1} - x_{i2} + x_{i3}, \\ D^{\alpha}x_{i3} = -p_2x_{i2}, \end{cases}$$

where p_1 and p_2 are system parameters and the $g(\cdot)$ is the piecewise linear function, which is described as

$$g(x_{i1}) = m_2x_{i1} + 0.5(m_1 - m_2)(|x_{i1} + 1| - |x_{i1} - 1|),$$

where $m_1 < m_2 < 0$ are two given constants, and further one gets the system matrix A and the nonlinear function $f(x_i)$ as

$$A = \begin{bmatrix} -p_1(1 + m_2) & p_1 & 0 \\ 1 & -1 & 1 \\ 0 & -p_2 & 0 \end{bmatrix},$$
$$f(x_i) = \begin{bmatrix} -0.5p_1(m_1 - m_2)(|x_{i1} + 1| - |x_{i1} - 1|) \\ 0 \\ 0 \end{bmatrix}.$$

Let $p_1 = 9.21$, $p_2 = 15.995$, $m_1 = -1.25$, $m_2 = -0.758$. Then it yields $\lambda_A = 16.5492$, $l = |(m_1 - m_2)p_1| = 4.5313$. In this chapter, we consider $\alpha = 0.97$, then the curve of the Mittag–Leffler function $E_{0.97}(x)$ is shown in Fig. 6.1. From Lemma 6.2 and Fig. 6.1, it implies $E_{0.97}(x) > 1$ when $x > 0$.

Example 6.1 Consider the MASs with fixed topology as shown in Fig. 6.2. From the topology in Fig. 6.2, the adjacency matrix, in-degree matrix, Laplacian matrix, and pinning matrix are obtained as follows:

$$\mathcal{A} = \begin{bmatrix} 0\,1\,0\,0 \\ 1\,0\,0\,0 \\ 0\,0\,0\,1 \\ 0\,0\,1\,0 \end{bmatrix}, D = \begin{bmatrix} 1\,0\,0\,0 \\ 0\,1\,0\,0 \\ 0\,0\,1\,0 \\ 0\,0\,0\,1 \end{bmatrix},$$
$$L = D - \mathcal{A} = \begin{bmatrix} 1 & -1 & 0 & 0 \\ -1 & 1 & 0 & 0 \\ 0 & 0 & 1 & -1 \\ 0 & 0 & -1 & 1 \end{bmatrix}, C = \begin{bmatrix} 1\,0\,0\,0 \\ 0\,0\,0\,0 \\ 0\,0\,0\,0 \\ 0\,0\,0\,1. \end{bmatrix}$$

Let $\xi = 1.01$, then from the condition (6.15) in Theorem 6.1, one gets the estimation of the impulsive interval as

$$\tau_k = t_k - t_{k-1} \leq -\frac{\ln(\lambda_M \xi)}{\lambda_A + 2l} = -\frac{\ln(1.01\lambda_k)}{25.6118}. \quad (6.29)$$

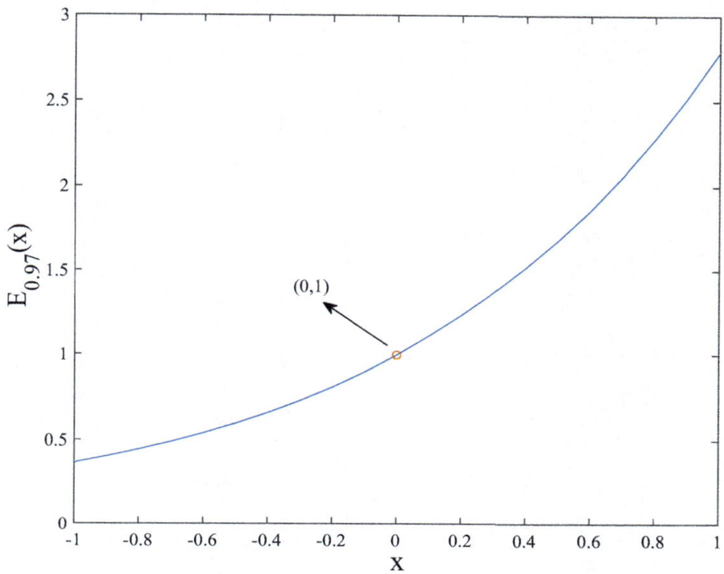

Fig. 6.1 The curve of Mittag–Leffler function $E_{0.97}(x)$

Fig. 6.2 Topology graph of the multi-agent systems in Example 6.1

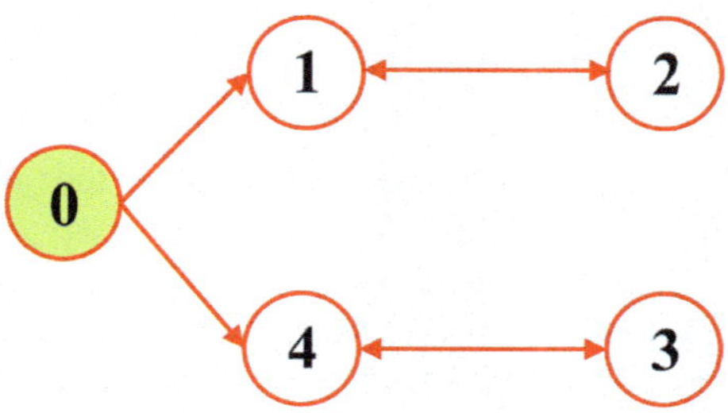

It should be noted that in (6.29), the gain error factor $\varphi(t_k)$ is needed for any $k \in \mathbb{N}^+$, which is not practical in real systems. Next, we adopt the results in Theorem 6.2 to further investigate the same consensus scheme. Firstly, we estimate the range of $b_k^* = b_k + \Delta b_k$ and get the lower and the upper bounds (ρ_1 and ρ_2). Figure 6.3 shows the allowable range of b_k^* under the condition $\lambda_k \in (0, 1)$, which implies $\rho_1 = -1$ and $\rho_2 = 0$. From Fig. 6.3, the consensus of MASs can be realized as long as $b_k^* \in (-1, 0)$. From (6.28), one can choose $m < 1$, then we choose $m = 0.1$ in this example. From (6.27), the selectable range of impulsive control gain is $b_k \in (-0.91, 0)$, then we select $b_k = -0.6$ in this example. Since $(1 + m)b_k < (1 + m\varphi(t_k))b_k < (1 - m)b_k$, one gets the range of b_k^* is $b_k^* \in (-0.66, -0.54)$. From Fig. 6.3, one obtains $\lambda_M = 0.6461$. From (6.26) in Theorem 6.2, one gets the impulsive interval can be chosen as

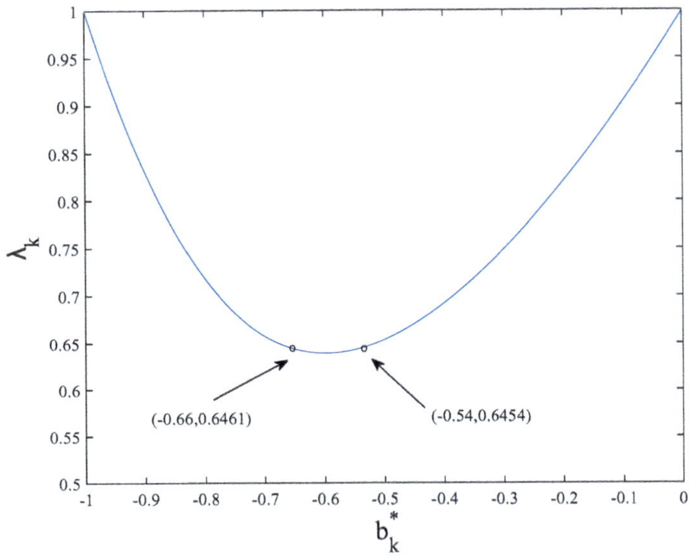

Fig. 6.3 The relation between λ_k and b_k^* in Example 6.1

$$\tau_k = t_k - t_{k-1} \leq -\frac{\ln(\lambda_M \xi)}{\lambda_A + 2l} = -\frac{\ln(1.01 * 0.6461)}{25.6118}. \tag{6.30}$$

In this example, we choose the initial values of each agent as $x_0 = [0.2, -0.1, 0.1]^T$, $x_1 = [0, 0.2, -0.3]^T$, $x_2 = [0, 0.2, -0.3]^T$, $x_3 = [-0.1, 0.1, 0.2]^T$, $x_4 = [-0.2, 0.3, 0.3]^T$. Figure 6.4 shows the relation between t_k and k in Example 6.1, which indicates that the impulsive instant t_k increases uniformly with the increase of k. Figure 6.5 shows the curve of impulsive interval in this example. Since λ_M is a fixed constant, from (6.30), one gets the obtained impulsive interval is also a invariant constant, which is shown in Fig. 6.5.

Figure 6.6 presents the global consensus error of MASs, which shows the consensus of MASs can be realized under the distributed impulsive control protocol with control gain error.

Example 6.2 For showing the effectiveness of the proposed protocol to other topology, in this example, we choose the topology $\mathcal{G}_2$ with spinning tree as presented in Fig. 6.7.

From the topology in Fig. 6.7, the adjacency matrix, in-degree matrix, Laplacian matrix, and pinning matrix are obtained as follows:

$$\mathcal{A} = \begin{bmatrix} 0\,1\,0\,1 \\ 1\,0\,1\,0 \\ 0\,1\,0\,1 \\ 1\,0\,1\,0 \end{bmatrix}, D = \begin{bmatrix} 2\,0\,0\,0 \\ 0\,2\,0\,0 \\ 0\,0\,2\,0 \\ 0\,0\,0\,2 \end{bmatrix},$$

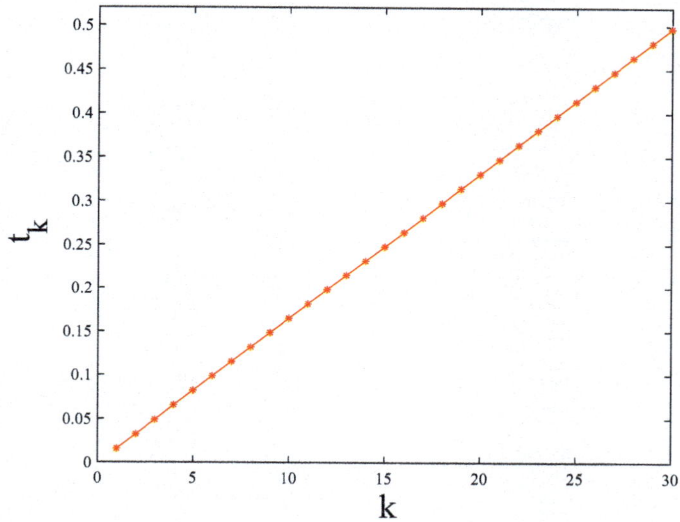

Fig. 6.4 The relation between t_k and k in Example 6.1

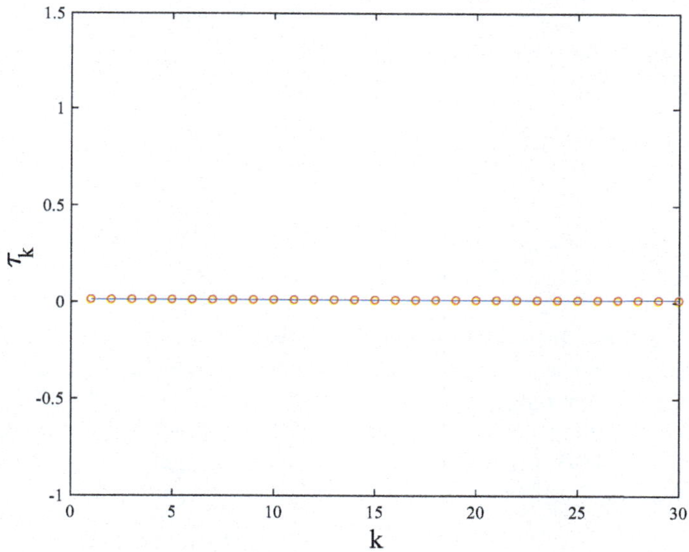

Fig. 6.5 Impulsive interval τ_k versus k in Example 6.1

$$L = D - \mathcal{A} = \begin{bmatrix} 2 & -1 & 0 & -1 \\ -1 & 2 & -1 & 0 \\ -1 & 0 & -1 & 2 \end{bmatrix}, C = \begin{bmatrix} 1\,0\,0\,0 \\ 0\,0\,0\,0 \\ 0\,0\,0\,0 \\ 0\,0\,0\,0. \end{bmatrix}$$

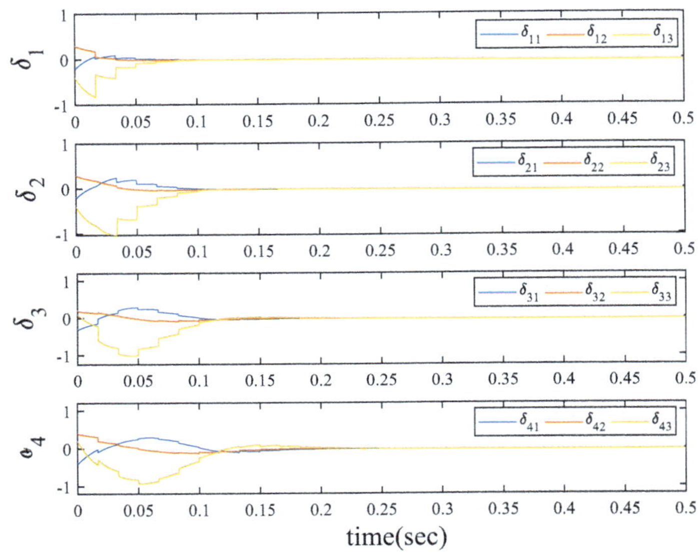

Fig. 6.6 The global consensus error in Example 6.1

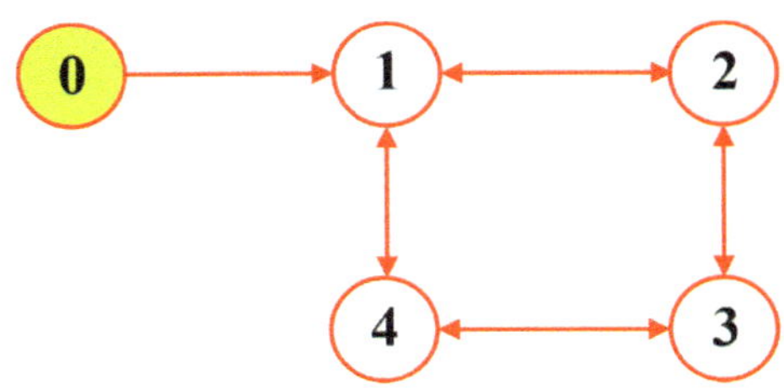

Fig. 6.7 Topology graph of the multi-agent systems in Example 6.2

Figure 6.8 shows the allowable range of b_k^* is $b_k^* \in (\rho_1, \rho_2) = (-0.46, 0)$. From (6.28), one can choose $m < 1$, then we also choose $m = 0.1$ in this example. From (6.27), the selectable range of impulsive control gain is $b_k \in (-0.418, 0)$, then we select $b_k = -0.4$ in this example, and get $b_k^* \in (-0.44, -0.36)$. From Fig. 6.8, one obtains $\lambda_M = 0.8703$. From (6.26) in Theorem 6.2, one gets the impulsive interval can be chosen as

$$\tau_k = t_k - t_{k-1} \leq -\frac{\ln(\lambda_k \xi)}{\lambda_A + 2l} = -\frac{\ln(1.01 * 0.8703)}{25.6118}. \tag{6.31}$$

The initial values of agents in MASs are the same as those in Example 6.1. Figure 6.9 presents the curve of t_k and k in Example 6.2, Fig. 6.10 shows the curve of impulsive interval in Example 6.2. Compared with Figs. 6.5 and 6.10, one obtains with the change of topology, the obtained impulsive interval will be different. It should be noted that the λ_M in this example is bigger than Example 6.1, which implies that

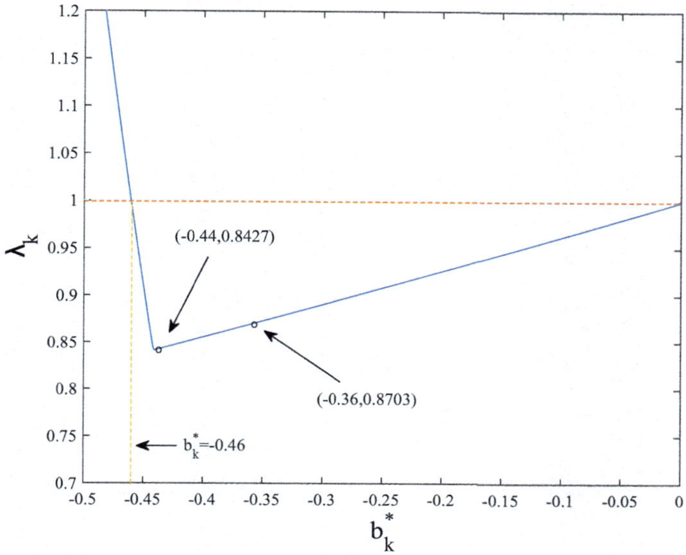

Fig. 6.8 The relation between λ_k and b_k^* in Example 6.2

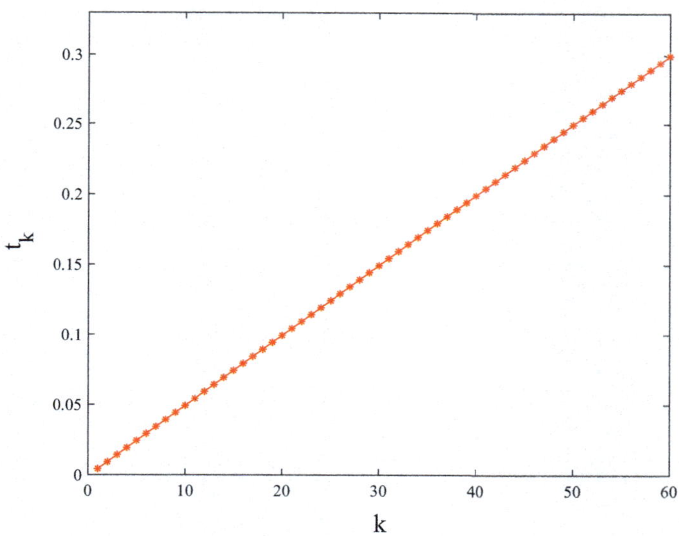

Fig. 6.9 The relation between t_k and k in Example 6.2

the effect of impulsive in this example is smaller than Example 6.1. Therefore, to verify that the results of this chapter can achieve the consensus of MASs, we present 60 impulses in this example.

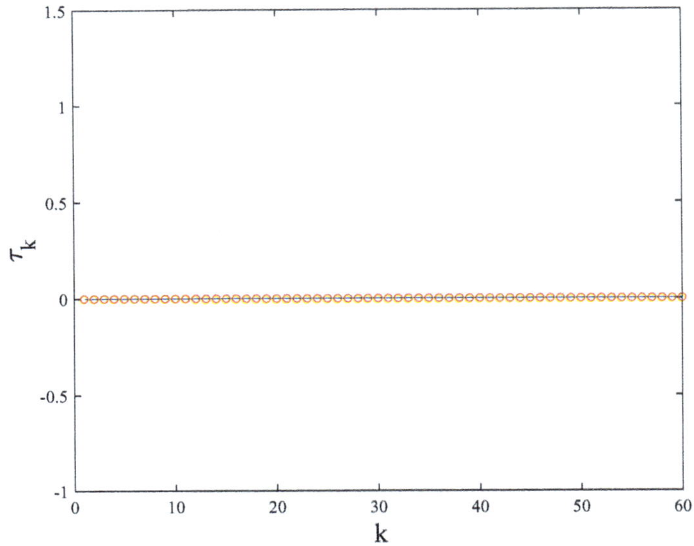

Fig. 6.10 Impulsive interval τ_k versus k in Example 6.2

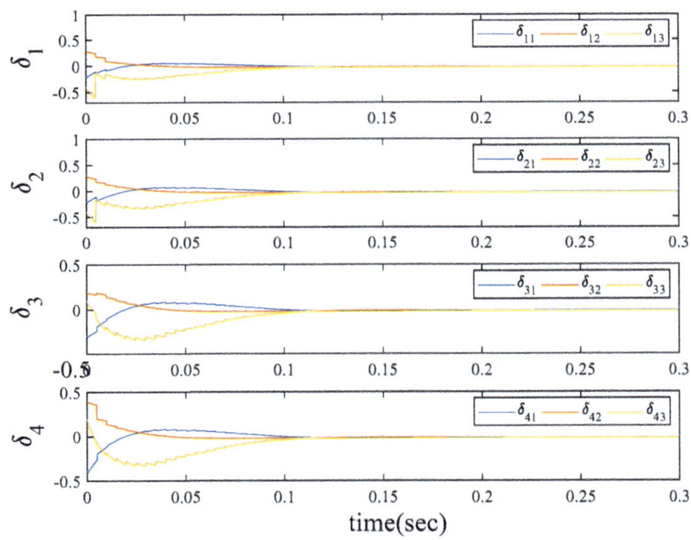

Fig. 6.11 The global consensus error in Example 6.2

The curve of global consensus error for agents in MASs is shown in Fig. 6.11, which also shows that the proposed impulsive control protocol is effective.

6.5 Summary

In this chapter, the consensus of fractional-order nonlinear MASs via distributed impulsive control strategy is studied. Some new sufficient conditions are derived for the consensus of MASs. In addition, this chapter considers the influence of control gain error on impulsive cooperative control and proposes a distributed impulsive control strategy. Then we improve the proposed method to enhance its practicability and feasibility. Finally, the correctness and effectiveness of the proposed method in this chapter are verified through two simulation examples with different topological structures.

References

Cao Y, Li Y, Ren W, Chen Y (2008) Distributed coordination algorithms for multiple fractional-order systems. In: Proceedings of the IEEE conference on decision and control, pp 2920–2925

Cao Y, Li Y, Ren W, Chen Y (2010) Distributed coordination of networked fractional-order systems. IEEE Trans Syst Man Cybern B Cybern 40(2):362–370

Cohen I, Golding I, Ron IG, Ben-Jacob E (2001) Biofluiddynamics of lubricating bacteria. Math Methods Appl Sci 24(17–18):1429–1468

Kozlovsky Y, Cohen I, Golding I, Ben-Jacob E (1999) Lubricating bacteria model for branching growth of bacterial colonies. Phys Rev E 59(6):7025–7035

Lin W, Peng S, Fu Z, Chen T, Gu Z (2022) Consensus of fractional-order multi-agent systems via event-triggered pinning impulsive control. Neurocomputing 494:409–417

Ma T, Zhao F (2014) Impulsive stabilization of a class of nonlinear system with bounded gain error. Chin Phys B 23(12):120504

Ma T, Zhang L, Zhao F, Gu Z, Yu X (2016a) Impulsive consensus of multi-agent nonlinear systems with control gain error. Neurocomputing 171:293–298

Ma T, Zhang L, Gu Z (2016b) Further studies on impulsive consensus of multi-agent nonlinear systems with control gain error. Neurocomputing 190:140–146

Narayanan G, Ali MS, Alsulami H, Stamov G, Stamova I, Ahmad B (2022) Impulsive security control for fractional-order delayed multi-agent systems with uncertain parameters and switching topology under DoS attack. Inf Sci 618:169–190

Podlubny I (1999) Fractional differential equations. Academic, New York, NY

Qing N, Yang Y, Luan X, Wan H (2024) Practical time-boundary consensus for fractional-order multi-agent systems under well-known and estimable topology. Appl Math Comput 464:128400

Sun F, Han Y, Zhu W, Kurths J (2024) Group consensus for fractional-order heterogeneous multi-agent systems under cooperation-competition networks with time delays. Commun Nonlinear Sci Numer Simul 133:107951

Xiao P, Li T, Ma T (2017a) Impulsive consensus of fractional-order multi-agent systems. In: Chinese control conference (CCC), pp 8644–8648

Xiao P, Zhang Z, Ma T (2017b) Impulsive consensus of uncertain multi-agent systems with control gain error. In: Proceedings - 2017 Chinese automation congress, CAC, pp 7757–7760

Yan X, Li K, Zhuang J, Yang C, Cao J (2024) Boundary control strategies for consensus of fractional-order multi-agent systems based on coupling PDE-ODEs. IEEE Trans Circuits Syst II Express Briefs 71(4):2179–2183

Yang J, Feckan M, Wang J (2023) Consensus of linear conformable fractional order multi-agent systems with impulsive control protocols. Asian J Control 25(1):314–324

Yu Z, Jiang H, Hu C (2015) Leader-following consensus of fractional-order multi-agent systems under fixed topology. Neurocomputing 149:613–620
Zhang L, Li T, Huang T, Huang J, Ma T (2017) Adaptive impulsive consensus of multi-agent systems with control gain error. In: Proceedings of the 29th Chinese Control and decision conference, CCDC, pp 4610–4615

Chapter 7
Distributed Impulsive Consensus of Multi-agent Systems Under Cyber Attacks

7.1 Introduction

In the actual complex environment, multi-agent systems (MASs) are vulnerable to malicious attackers. Denial-of-Service (DoS) attacks refer to a type of attack that destroys some or all components of the network system (Ding et al. 2022; Bu et al. 2023; Zhang et al. 2023). At present, there are some research results on impulsive coordination control under DoS attacks. For example, Wang et al. (2019) studied impulsive coordination of systems under DoS attacks. However, when analyzing the impulsive instants, only the worst case was considered, which will greatly increase the conservatism of the obtained results. Ye et al. (2019) investigated the impulsive quasi-synchronization control subjective to DoS attacks. However, the impulses without attacks are not guaranteed to reduce the consensus error. Guo et al. (2023) studied the impulsive coordination problem of directed graphs under DoS attacks. However, the designed controller method needs to measure the state and error of the systems in the continuous interval, which causes an increase in information transmission cost.

Another common type of cyber attack is a deception attack, which means that the attackers can access in system, manipulate transmission data, and control commands (Li and Zhou 2023). In practice, deception attacks are harder to detect than DoS attacks and seriously affect data integrity. Recently, there are some research works concerning the impulsive consensus of MASs with deception attacks. He et al. (2018) firstly considered the impulsive consensus of MASs with deception attacks. Wen et al. (2020) achieved the impulsive synchronization for delayed uncertain MASs with deception attacks. He et al. (2020) researched the consensus of Lipchitz MASs with deception attacks via impulsive protocol. Liu et al. (2023) achieved the impulsive synchronization of MASs with deception attacks and switching topologies. Li et al. (2022) realized the adaptive impulsive security consensus of MASs under randomly occurring nonlinearities and deception attacks. Chen et al. (2023) realized the synchronization of Lipschitz MASs under deception attacks by using guaranteed performance impulsive control.

Z. Zhang et al., *Distributed Impulsive Coordination of Multi-Agent Systems*, Intelligent Control and Learning Systems 15, https://doi.org/10.1007/978-981-97-6262-0_7

In this chapter, the distributed impulsive controllers under cyber attacks are designed. Firstly, according to the recovery mechanism of DoS attacks, the minimum interval between two adjacent DoS attacks is obtained. Based on the Lyapunov stability theory and impulsive control theory, sufficient conditions for control parameters under DoS attacks are derived. Different from the existing research results, the method proposed in this chapter can flexibly deal with the communication interruption and topology switching caused by DoS attacks. Next, this chapter considers the deception attacks in the information transmission process of the controller-to-actuator channel and models the deception attacks as the Bernoulli distribution to derive sufficient conditions for achieving the coordination of MASs under deception attacks. Compared with the current results, the control strategy designed in this chapter considers the attack probability of different agents is different and can avoid the unreasonable situation that the global consensus error of MASs is reduced when the continuous interval is not controlled. In addition, this chapter analyzes the influence of different control parameters on the control effect under deception attacks, especially discusses the effect of exponential convergence rate, which is often overlooked in the existing literature. Finally, the effectiveness of the impulsive control strategy designed in this chapter is verified by simulation.

7.2 Problem Description

Consider the model of the follower agent as

$$\dot{x}_i(t) = Ax_i(t) + f(x_i(t)) + u_i(t), \tag{7.1}$$

where $x_i(t) \in \mathbb{R}^n$ is the state of agent i, $(i = 1, 2, \ldots N)$, $f : \mathbb{R}^n \to \mathbb{R}^n$ is a continuous nonlinear function of agent i, which satisfies the Lipschitz condition. $A \in \mathbb{R}^{n\times n}$ is a known system matrix.

The model of leader is

$$\dot{x}_0(t) = Ax_0(t) + f(x_0(t)), \tag{7.2}$$

where $x_0(t) \in \mathbb{R}^n$ is the state of leader node.

Definition 7.1 The global synchronization error is

$$\delta_i(t) = x_i(t) - x_0(t), \quad \delta_i(t) \in \mathbb{R}^n. \tag{7.3}$$

Definition 7.2 The local neighborhood synchronization error for agent i is

$$e_i(t) = \sum_{j\in N_i} a_{ij}(x_i(t) - x_j(t)) + c_i(x_i(t) - x_0(t)). \tag{7.4}$$

For the consensus problem under DoS attacks, this chapter considers the situation where communication between some or all edges of MASs will be interrupted under DoS attacks, resulting in a change in the communication topology of the MASs. It should be noted that the energy of DoS attacks is limited and cannot be continuously attacked, and MASs have recovery mechanisms that can restore communication topology after the attack. Attackers can only attack the systems for a period of time, and after the attack ends, the attacker will enter idle time to store energy for the next attack. Therefore, there is a minimum interval between two adjacent DoS attacks (i.e., from the end of the last DoS attack to the start of the current attack).

For the consensus problem under deception attacks, assume the attacks in controller-to-actuator communication channel. A random variable $\gamma_i(t)$ associated with the Bernoulli distribution is considered in this chapter. $\gamma_i(t) = 1$ means that the attacker implements deception attacks, otherwise, $\gamma_i(t) = 0$. Define $q_i(t) \in \mathbb{R}^n$ as attack signals from attackers. Denote $Q(t) = \left[q_1^T(t), \ldots, q_N^T(t)\right]^T$.

7.3 Main Results

7.3.1 *Impulsive Consensus of Multi-agent Systems Under DoS Attacks*

The distributed impulsive controller under DoS attacks is designed as

$$u_i(t) = \sum_{k=1}^{\infty} b\left(\sum_{j\in N_i} a_{ij}(m)\left(x_i(t) - x_j(t)\right) + c_i(m)\left(x_i(t) - x_0(t)\right)\right)\delta^*(t - t_k), \tag{7.5}$$

where $i = 1, 2, \ldots, N$, b is the impulsive control gain, the impulsive sequence $\{t_k\}$ satisfying $0 \le t_0 < t_1 < t_2 < \cdots < t_{k-1} < t_k < \cdots$, $\lim_{k\to\infty} t_k = \infty$, $\delta(t)$ is the Dirac delta function, which satisfies $\delta(t) = 0$ for $t \neq 0$ and $\int_{-\infty}^{+\infty}\delta(t)\,dt = 1$. m represents the different topology of MASs, $m = 0$ means the MASs not under DoS attacks at $t = t_k$, $m = 1, 2, 3, \ldots$ represents the topology under different DoS attacks.

Substituting impulsive controller (7.5) into the system model of the follower (7.1), one gets

$$\begin{cases} \dot{x}_i(t) = Ax_i(t) + f(x_i(t)), \quad t \neq t_k, \\ \Delta x_i(t_k) = x_i(t_k^+) - x_i(t_k^-) \\ \quad = b\left(\sum_{j\in N_i} a_{ij}(m)\left(x_i(t_k) - x_j(t_k)\right) + c_i(m)\left(x_i(t_k) - x_0(t_k)\right)\right) \\ x_i(t_0^+) = x_i(t_0), \quad t_0 \ge 0, \quad i = 1, 2, \cdots, N, \quad k \in \mathbb{N}^+, \end{cases} \tag{7.6}$$

where $\Delta x_i(t_k)$ is the jump of the state of node i at impulse instant t_k, and satisfies left continuous at t_k, i.e., $x_i(t_k) = x_i(t_k^-)$.

From (7.2) and (7.6), one obtains the global consensus error system as

$$\begin{cases} \dot{\delta}_i(t) = A\delta_i(t) + f(x_i(t)) - f(x_0(t)), \quad t \neq t_k, \\ \Delta\delta_i(t_k) = b\left(\sum_{j\in N_i} a_{ij}(m)\left(x_i(t_k) - x_j(t_k)\right) + c_i(m)\left(x_i(t_k) - x_0(t_k)\right)\right). \end{cases} \tag{7.7}$$

By using the properties of the Kronecker product, the system (7.7) is rewritten as

$$\begin{cases} \dot{\delta}(t) = (I_N \otimes A)\delta(t) + \bar{F}(x(t), \bar{x}_0(t)), \quad t \neq t_k, \\ \Delta\delta(t_k) = b\left((L(m) + C(m)) \otimes I_n\right)\delta(t_k), \end{cases} \tag{7.8}$$

where $x(t) = [x_1^T(t), x_2^T(t), \ldots, x_N^T(t)]^T \in \mathbb{R}^{Nn}$, $\delta(t) = [\delta_1^T(t), \delta_2^T(t), \ldots, \delta_N^T(t)]^T \in \mathbb{R}^{Nn}$, $\bar{f}(x(t), \bar{x}_0(t)) = [f^T(x_1(t)) - f^T(x_0(t)), \ldots, f^T(x_N(t)) - f^T(x_0(t))]^T \in \mathbb{R}^{Nn}$, $\bar{x}_0(t) = 1_N \otimes x_0(t) \in \mathbb{R}^{Nn}$, $L(m)$ and $C(m)$ represent the Laplacian matrix and pinning matrix of the corresponding topological structure m, respectively, $m = 0$ means that the MASs not under DoS attacks, and $m \neq 0$ means that the MASs suffering from DoS attacks.

The control objective of this subsection is to apply distributed impulsive control to all following nodes in MASs under DoS attacks and design appropriate strategies under DoS attacks, so that the states of all followers can realize consensus of the leader with any initial condition, i.e.,

$$\lim_{t\to\infty} \delta(t) = 0. \tag{7.9}$$

In order to eliminate the impact of DoS attacks and achieve consensus in MASs, this subsection divides the control process into three colored windows, as shown in Fig. 7.1. The green, red, and blue areas in Fig. 7.1 represent the ranges of the normal window, attack window, and adjustment window for the pth period, respectively.

The characteristics and functions of the three color windows are described as follows:

(1) Normal window (green window): The length of the green window is the interval after the last blue window and before the next DoS attack. Since the start of the next DoS attack cannot be predicted, it is difficult to obtain the length of the green window in advance. It should be noted that if the next DoS attack occurs at the end of the blue window, the green window will not exist.

(2) Attack window (red window): The length of the red window indicates the duration of the DoS attack. Notice that the start instant and the duration of each DoS attack cannot be predicted in advance. This is in line with the actual situation, as attackers do not give any in advance when the attack will start or the duration of the attack, and the timing and duration of the attack are often irregular.

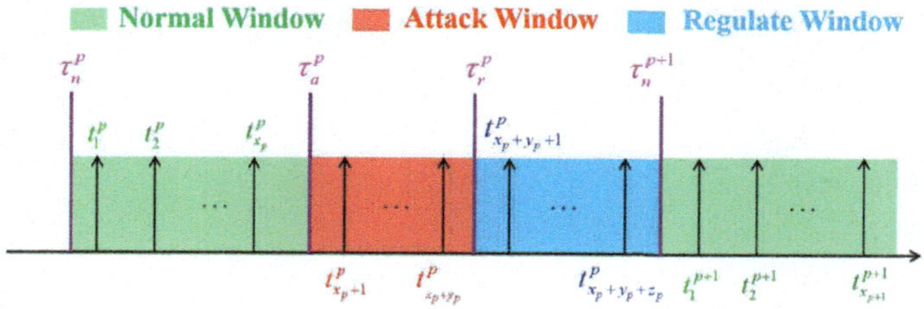

Fig. 7.1 The diagram of DoS attack and impulsive instants

(3) Regulate window (blue window): The length of the regulate window is equal to the minimum interval between two adjacent attacks. Among the three different colored windows, only the length of the blue window is known. To eliminate the impact of DoS attacks, it is necessary to apply an appropriate number of impulsive control in the blue window to ensure that the global bipartite consensus error of the MASs decreases after a complete period (that is, the normal window, attack window, and regulate window).

The following theorem is given to realize the impulsive consensus of MASs under DoS attacks.

Theorem 7.1 *Assume the MASs satisfy Assumptions 1.1 and 1.2, if*

$$(\lambda_A + 2l)(\tau_n^{p+1} - \tau_n^p) + \ln(\lambda^{x_p+z_p}\lambda_m^{y_p}\xi) \leq 0, \tag{7.10}$$

where $\xi > 1$, λ_A, λ *and* λ_m *are the maximum eigenvalue of* $A + A^T$, $(b(L(0) + C(0)) + I_N)^T\, (b(L(0) + C(0)) + I_N)$ *and* $(b(L(m) + C(m)) + I_N)^T$ $(b(L(m) + C(m)) + I_N)$ $(m \neq 0)$, *respectively. Then the consensus of MASs will be realized under the control strategy designed in this section.*

Proof Choose the Lyapunov function as $V(t) = \delta^T(t)\delta(t)$, when $t \in (t_{k-1}, t_k]$, one has

$$\begin{aligned} D^+V(t) &= \dot{\delta}^T(t)\delta(t) + \delta^T(t)\dot{\delta}(t) \\ &= \delta^T(t)(I_N \otimes (A + A^T))\delta(t) + 2\delta^T(t)\bar{f}(x(t), \bar{x}_0(t)) \\ &\leq (\lambda_A + 2l)\delta^T(t)\delta(t) \\ &= (\lambda_A + 2l)V(t), \end{aligned} \tag{7.11}$$

which leads to

$$V(t) \leq V(t_{k-1}^+)\exp((\lambda_A + 2l)(t - t_{k-1})). \tag{7.12}$$

When $t = t_k$ (t_k in green or blue window),

$$
\begin{aligned}
& V(t_k^+) \\
& = \delta^T(t_k^+)\delta(t_k^+) \\
& = \delta^T(t_k)\left(\left((b\,(L(0)+C(0))+I_N)^T\,(b\,(L(0)+C(0))+I_N)\right)\otimes I_n\right)\delta(t_k) \\
& \le \lambda\delta^T(t_k)\delta(t_k) \\
& = \lambda V(t_k).
\end{aligned}
\tag{7.13}
$$

When t_k is in red window,

$$
\begin{aligned}
& V(t_k^+) \\
& = \delta^T(t_k^+)\delta(t_k^+) \\
& = \delta^T(t_k)\left(\left((b\,(L(m)+C(m))+I_N)^T\,(b\,(L(m)+C(m))+I_N)\right)\otimes I_n\right)\delta(t_k) \\
& \le \lambda_m\delta^T(t_k)\delta(t_k) \\
& = \lambda_m V(t_k).
\end{aligned}
\tag{7.14}
$$

When $t \in (t_0, t_1)$, from (7.12),

$$
V(t) \le V(t_0)\exp((\lambda_A + 2l)(t - t_0)), \tag{7.15}
$$

which leads to

$$
V(t_1) \le V(t_0)\exp((\lambda_A + 2l)(t_1 - t_0)). \tag{7.16}
$$

When $t \in (t_1, t_2]$, from (7.12), (7.13), and (7.16)

$$
\begin{aligned}
V(t) & \le V(t_1^+)\exp((\lambda_A + 2l)(t - t_1)) \\
& \le \lambda V(t_1)\exp((\lambda_A + 2l)(t - t_1)) \\
& \le \lambda V(t_0)\exp((\lambda_A + 2l)(t_1 - t_0))\exp((\lambda_A + 2l)(t - t_1)) \\
& = \lambda V(t_0)\exp((\lambda_A + 2l)(t - t_0)).
\end{aligned}
\tag{7.17}
$$

When $t = \tau_a^1$, the MASs have just passed the first green window,

$$
\begin{aligned}
V(t) & \le V(t_{x_1}^{1+})\exp((\lambda_A + 2l)(\tau_a^1 - t_{x_1}^p)) \\
& \le \lambda^{x_1} V(t_0)\exp((\lambda_A + 2l)(\tau_a^1 - t_0)).
\end{aligned}
\tag{7.18}
$$

When $t = \tau_r^1$, the MASs have just passed the first green and red window,

$$
\begin{aligned}
V(\tau_r^1) & \le V(t_{x_1+y_1}^{1+})\exp((\lambda_A + 2l)(\tau_r^1 - t_{x_1+y_1}^1)) \\
& \le \lambda^{x_1} V(t_0)\exp((\lambda_A + 2l)(\tau_a^1 - t_0))\lambda_m^{y_1}\exp((\lambda_A + 2l)(\tau_r^1 - \tau_a^1)).
\end{aligned}
\tag{7.19}
$$

When $t = \tau_n^2$, the MASs have just passed through the first green, red, and blue windows (i.e., have just gone through the first period),

$$
\begin{aligned}
V(\tau_n^2) &\leq V(t_{x_1+y_1+z_1}^{1+})\exp((\lambda_A+2l)(\tau_n^2-t_{x_1+y_1+z_1})) \\
&\leq \lambda^{x_1}V(t_0)\exp((\lambda_A+2l)(\tau_a^1-t_0))\lambda_m^{y_1} \\
&\quad \exp((\lambda_A+2l)(\tau_r^1-\tau_a^1))\lambda^{z_1}\exp((\lambda_A+2l)(\tau_n^2-\tau_r^1)) \\
&= \lambda^{x_1+z_1}\lambda_m^{y_1}V(t_0)\exp((\lambda_A+2l)(\tau_n^2-t_0)).
\end{aligned}
\tag{7.20}
$$

Similarly, when $t=\tau_n^3$, the MASs have passed through the first and second period,

$$
\begin{aligned}
V(\tau_n^3) &\leq V(t_{x_1+y_1+z_1}^{1+})\exp((\lambda_A+2l)(\tau_n^2-t_{x_1+y_1+z_1})) \\
&\leq \lambda^{x_1}V(t_0)\exp((\lambda_A+2l)(\tau_a^1-t_0))\lambda_m^{y_1}\exp((\lambda_A+2l)(\tau_r^1-\tau_a^1)) \\
&\quad \lambda^{z_1}\exp((\lambda_A+2l)(\tau_n^2-\tau_r^1)) \\
&= \lambda^{x_1+z_1}\lambda_m^{y_1}V(t_0)\exp((\lambda_A+2l)(\tau_n^2-t_0)).
\end{aligned}
\tag{7.21}
$$

In general, when $t=\tau_n^{p+1}$, the MASs have gone through p periods

$$
\begin{aligned}
V(\tau_n^{p+1}) &\leq V(t_{x_p+y_p+z_p}^{p+})\exp((\lambda_A+2l)(\tau_n^{p+1}-t_{x_p+y_p+z_p}^{p})) \\
&\leq \lambda^{x_p}V(\tau_n^p)\exp((\lambda_A+2l)(\tau_a^p-\tau_n^p))\lambda_m^{y_p}\exp((\lambda_A+2l)(\tau_r^p-\tau_a^p)) \\
&\quad \lambda^{z_p}\exp((\lambda_A+2l)(\tau_n^{p+1}-\tau_r^p)) \\
&= \lambda^{x_p+z_p}\lambda_m^{y_p}V(\tau_{p-1}^3)\exp((\lambda_A+2l)(\tau_n^{p+1}-\tau_n^p)) \\
&= V(t_0)\lambda^{x_p+z_p}\lambda_m^{y_p}\exp((\lambda_A+2l)(\tau_n^{p+1}-\tau_n^p))\cdots \\
&\quad \lambda^{x_1+z_1}\lambda_m^{y_1}\exp((\lambda_A+2l)(\tau_n^2-\tau_n^1)) \\
&\leq \frac{1}{\xi^p}V(t_0).
\end{aligned}
\tag{7.22}
$$

Since $V(t_0)$ is a bounded constant, and $1/\xi^p\to 0$ when $p\to\infty$, then $\lim_{p\to\infty}V(\tau_n^{p+1})=0$.

When $t\in(\tau_n^{p+1},t_1^{p+1}]$, (i.e., from the left endpoint of the pth green window to the first impulsive instant in the $(p+1)$th green window), from (7.12),

$$
V(t)\leq V(\tau_n^{p+1})\exp((\lambda_A+2l)(t-\tau_n^{p+1})). \tag{7.23}
$$

When $t\in(t_1^{p+1},t_2^{p+1}]$, (i.e., from the first impulsive instant in the $(p+1)$th green window to the second impulsive instant in the $(p+1)$th green window), from (7.12) and (7.13),

$$
\begin{aligned}
V(t) &\leq \lambda V(t_1^{p+1})\exp((\lambda_A+2l)(t-t_1^{p+1})) \\
&\leq \lambda V(\tau_n^{p+1})\exp((\lambda_A+2l)(t-\tau_n^{p+1})).
\end{aligned}
\tag{7.24}
$$

When $t\in(t_{x_{p+1}}^{p+1},\tau_a^{p+1}]$, (i.e., from the last impulsive instant in $(p+1)$th green window to the end of the $(p+1)$th green window),

$$\begin{aligned} V(t) &\le V(t_{x_{p+1}}^{p+1}) \exp((\lambda_A + 2l)(\tau_a^{p+1} - t_{x_{p+1}}^{p+1})) \\ &\le \lambda^{x_{p+1}} V(\tau_n^{p+1}) \exp((\lambda_A + 2l)(\tau_a^{p+1} - \tau_n^{p+1})). \end{aligned} \tag{7.25}$$

Since λ, x_{p+1} and $\exp((\lambda_A + 2l)(\tau_a^{p+1} - \tau_n^{p+1}))$ are bounded, $\lim_{p\to\infty} V(\tau_n^{p+1}) = 0$, then from (7.23) to (7.25), when $t \in (\tau_n^{p+1}, \tau_a^{p+1}]$ (i.e., in the $(p+1)$th green window), $\lim_{t\to\infty} V(t) = 0$.

When $t \in (\tau_a^{p+1}, t_{x_{p+1}+1}^{p+1}]$, (i.e., from the left endpoint of the pth red window to the first impulsive instant in the $(p+1)$th red window), from (7.12),

$$V(t) \le V(\tau_a^{p+1}) \exp((\lambda_A + 2l)(t - \tau_a^{p+1})). \tag{7.26}$$

When $t \in (t_{x_{p+1}+1}^{p+1}, t_{x_{p+1}+2}^{p+1}]$, (i.e., from the first impulsive instant in the $(p+1)$th red window to the second impulsive instant in the $(p+1)$th red window), from (7.12) and (7.14),

$$\begin{aligned} V(t) &\le \lambda_m V(t_{x_{p+1}+1}^{p+1}) \exp((\lambda_A + 2l)(t - t_{x_{p+1}+1}^{p+1})) \\ &\le \lambda_m V(\tau_n^{p+1}) \exp((\lambda_A + 2l)(t - \tau_n^{p+1})). \end{aligned} \tag{7.27}$$

When $t \in (t_{x_{p+1}+y_{p+1}}^{p+1}, \tau_r^{p+1}]$, (i.e., from the last impulsive instant in $(p+1)$th red window to the end of the $(p+1)$th red window),

$$\begin{aligned} V(t) &\le V(t_{x_{p+1}+y_{p+1}}^{p+1}) \exp((\lambda_A + 2l)(\tau_r^{p+1} - t_{x_{p+1}+y_{p+1}}^{p+1})) \\ &\le \lambda_m^{y_{p+1}} V(\tau_a^{p+1}) \exp((\lambda_A + 2l)(\tau_r^{p+1} - \tau_a^{p+1})). \end{aligned} \tag{7.28}$$

Since λ_m, y_{p+1} and $\exp((\lambda_A + 2l)(\tau_r^{p+1} - \tau_a^{p+1}))$ are bounded, $\lim_{p\to\infty} V(\tau_a^{p+1}) = 0$, then from (7.26) to (7.28), when $t \in (\tau_a^{p+1}, \tau_r^{p+1}]$ (i.e., in the $(p+1)$th red window), $\lim_{t\to\infty} V(t) = 0$.

When $t \in (\tau_r^{p+1}, t_{x_{p+1}+y_{p+1}+1}^{p+1}]$, (i.e., from the left endpoint of the pth blue window to the first impulsive instant in the $(p+1)$th blue window), from (7.12),

$$V(t) \le V(\tau_r^{p+1}) \exp((\lambda_A + 2l)(t - \tau_r^{p+1})). \tag{7.29}$$

When $t \in (t_{x_{p+1}+y_{p+1}+1}^{p+1}, t_{x_{p+1}+y_{p+1}+2}^{p+1}]$, (i.e., from the first impulsive instant in the $(p+1)$th blue window to the second impulsive instant in the $(p+1)$th blue window), from (7.12) and (7.13),

$$\begin{aligned} V(t) &\le \lambda V(t_{x_{p+1}+y_{p+1}+1}^{p+1}) \exp((\lambda_A + 2l)(t - t_{x_{p+1}+y_{p+1}+1}^{p+1})) \\ &\le \lambda V(\tau_r^{p+1}) \exp((\lambda_A + 2l)(t - \tau_r^{p+1})). \end{aligned} \tag{7.30}$$

When $t \in (t_{x_{p+1}+y_{p+1}+z_{p+1}}^{p+1}, \tau_n^{p+2}]$, (i.e., from the last impulsive instant in $(p+1)$th blue window to the end of the $(p+1)$th blue window),

$$
\begin{aligned}
V(t) &\le V(t^{p+1}_{x_{p+1}+y_{p+1}+z_{p+1}})\exp((\lambda_A+2l)(\tau_n^{p+2}-t^{p+1}_{x_{p+1}+y_{p+1}+z_{p+1}}))\\
&\le \lambda^{z_{p+1}}V(\tau_a^{p+1})\exp((\lambda_A+2l)(\tau_n^{p+2}-\tau_r^{p+1})).
\end{aligned}
\tag{7.31}
$$

Since λ, z_{p+1} and $\exp((\lambda_A+2l)(\tau_a^{p+1}-\tau_n^{p+1}))$ are bounded, and $\lim_{p\to\infty} V(\tau_n^{p+1}) = 0$, then from (7.29) to (7.31), when $t \in (\tau_r^{p+1}, \tau_n^{p+2}]$ (i.e., in the $(p+1)$th blue window), $\lim_{t\to\infty} V(t) = 0$.

Based on the analysis from (7.23) to (7.31), when $t \in (\tau_n^{p+1}, \tau_n^{p+2}]$ (i.e., in $(p+1)$th period), since $\lim_{p\to\infty} V(\tau_n^{p+1}) = 0$, then $\lim_{t\to\infty} V(t) = 0$. Moreover, since $\lim_{p\to\infty} V(\tau_n^{p+2}) = 0$, then $\lim_{t\to\infty} V(t) = 0$, which means that the global consensus error of MASs will become 0 when $t \to \infty$. This completes the proof.

Remark 7.1 The further explanation of Theorem 7.1 is as follows:

(1) In each period, only the length of the blue window is fixed, which is equal to the minimum interval between adjacent DoS attacks. However, because the beginning and duration of DoS attacks are uncertain and unpredictable, the length of the red and green windows cannot be predetermined. In addition, due to the inability to determine the corresponding topology of MASs in advance under DoS attacks, it is unrealistic to calculate the number and interval of impulses in the red and green windows in advance. Therefore, Theorem 7.1 is only used to calculate the required number of impulses in the blue window.

(2) In (7.10), $\tau_n^{p+1} - \tau_n^p$ is the length of pth period. Although the length of the green and red windows cannot be determined in advance, the duration of the two windows can be measured. Therefore, the value of $\tau_n^{p+1} - \tau_n^p$ can be obtained before calculating the required number of impulses in the blue window.

(3) In the red and green windows, the choice of impulsive interval is arbitrary, but we need to preset the number of impulses in the two windows, which are x_p and y_p, respectively.

(4) In (7.10), λ_A and l are determined by the system parameters, and the variable ξ can be chosen to be a little bigger than 1.

Based on the above discussion, in (7.10), only the number of impulses z_p in blue window needs to be calculated. From (7.10), one gets $z_p \ge \frac{-(\lambda_A+2l)(\tau_n^{p+1}-\tau_n^p)-ln(\xi\lambda^{x_p}\lambda_m^{y_p})}{\ln\lambda}$, then one can choose

$$
z_p = \left\lceil \frac{-(\lambda_A+2l)(\tau_n^{p+1}-\tau_n^p)-\ln(\xi\lambda^{x_p}\lambda_m^{y_p})}{\ln\lambda} \right\rceil. \tag{7.32}
$$

Remark 7.2 From (7.22), one obtains $\lim_{p\to\infty} V(\tau_n^{p+1}) = 0$. However, this can only indicate that when $p \to \infty$, $V(t)$ can only approach zero at the beginning of each green window, and it cannot be concluded that when considering continuous time, $V(t) \to 0$. Therefore, in the proof process, this section supplemented the analysis of all time periods in the normal window, attack window, and adjustment window of the

pth period when $\lim_{p\to\infty} V(\tau_n^{p+1}) = 0$ is satisfied, and obtains when $t \in (\tau_n^{p+1}, \tau_n^{p+2}]$ (i.e., in pth period), $\lim_{t\to\infty} V(t) = 0$. Furthermore, it can be concluded that when $p \to \infty$, $V(t) \to 0$ is satisfied at any time during the pth period. That is to say, the analysis of Eqs. (7.23) to (7.31) is an extension from a single instant to a time period to demonstrate that when considering continuous time $t \to \infty$, $V(t) \to 0$.

Remark 7.3 The parameter $(\lambda_A + 2l)$ is considered nonnegative in this section. In fact, if $(\lambda_A + 2l) < 0$, from (7.13), one obtains $V(t)$ can converge to zero exponentially when $t \to \infty$, which means that the consensus can be achieved without any control input, but this is difficult to achieve in practice. Therefore, in this section, $(\lambda_A + 2l) \geq 0$ is considered, which means that the global consensus error of the system will increase in continuous intervals which are not controlled, and the consensus can only be achieved by exerting impulsive control.

Remark 7.4 Notice that the impulsive control may have a negative effect during DoS attacks (i.e., the situation of $\lambda_m > 1$). After experiencing an impulsive control, it can be inferred from Eq. (7.14) that the global consensus error of MASs may increase. If the impact of DoS attacks cannot be eliminated in the blue window, the control objective cannot be achieved. Therefore, it is necessary to determine the number of impulses required in the blue window based on (7.10) in Theorem 7.1 and the algorithm given in Remark 7.1. Only when enough impulses are exerted in the blue window can realize the consensus of MASs under DoS attacks.

Remark 7.5 Compared with the variable impulsive and intermittent impulsive control methods in Chap. 2, it can be seen that the analysis method in this section has similarities with the variable impulsive control methods. The first similarity is that the control process is divided into multiple time windows, and multiple impulses need to be exerted in each time window. The second similarity is that the instant of each impulsive control need not be fixed and can be exerted at any time in the control window. However, compared to the variable and intermittent impulsive control discussed in Chap. 2, the problem studied in this section has the following differences and difficulties:

(1) In variable and intermittent impulsive control schemes, the topology of the MASs remains unchanged and needs to always satisfy the assumption that it contains a directed spanning tree. When the MASs is attacked, the attacker may disrupt the information transmission of any edge in the system, and the topology may no longer satisfy the directed spanning tree.

(2) In both variable and intermittent impulsive control schemes, all impulses play a role in promoting consensus, that is, by selecting the appropriate impulsive control gain, $\lambda_k \in (0, 1)$ can be achieved. When the MASs under DoS attacks, due to the possibility that the system may no longer satisfy the conditions for a directed spanning tree, λ_m in (7.14) may be greater than 1. This means that when the MASs under DoS attacks, the global consensus error of the system may increase after impulsive instants in red window, so enough impulsive control needs to be exerted in the adjustment window to eliminate the adverse effects of DoS attacks.

(3) In the variable impulsive control scheme, the end time of all windows can be set by the user themselves, and in the intermittent impulsive control scheme, the start and end times of each control window can also be set by the users themselves. However, in the case of DoS attacks, as the time of the attack is unpredictable, the beginning and the end instants of the red window in Fig. 7.1 (i.e., the time when DoS attacks occur and end) cannot be determined by the user themselves. Under DoS attacks, only the interval of the blue window (the minimum interval between adjacent DoS attacks) is known.

In summary, this section considers that when MASs are suffered from DoS attacks, the communication topology may not satisfy a directed spanning tree, and the impulses may not have a positive effect. Moreover, the start and end times of the attack cannot be predicted in advance, which makes the design difficulty of the control method for resisting DoS attacks in this chapter greater than that of the variable and intermittent impulsive methods in Chap. 2.

Remark 7.6 It is a necessary assumption in this chapter that no attacks occur in the regulate window, during this period the impact of DoS attacks needs to be eliminated by applying enough impulsive control. In fact, if DoS attacks occur in the blue window, the next attack will occur when the impact of the last DoS attack has not been eliminated. If the time of DoS attack in the blue window is known in advance, enough impulses can be exerted before the DoS attack to eliminate the impact of the last DoS attack. However, the time of DoS attack cannot be known in advance, so it is difficult to exert enough impulses before DoS attack, and the complexity of the problem is far beyond the scope of the current work.

Remark 7.7 From the proof of Theorem 7.1, it is obvious that when $t = t_k$ (t_k is in the red window), $V(t_k^+) \leq \lambda_m V(t_k)$. To save control resources, the impulsive control should not be exerted in the red window when $\lambda_m \geq 1$. Motivated by the above discussion, the following corollary is given.

Corollary 7.1 *Under the Assumptions 1.1 and 1.2, if there exists a constant $\xi > 1$ such that*

$$(\lambda_A + 2l)(\tau_n^{p+1} - \tau_n^p) + \ln(\lambda^{x_p+z_p}\lambda_m^{y_p}\xi) \leq 0, \quad \lambda_m < 1 \tag{7.33}$$

$$(\lambda_A + 2l)(\tau_n^{p+1} - \tau_n^p) + \ln(\lambda^{x_p+z_p}\xi) \leq 0, \quad \lambda_m \geq 1 \tag{7.34}$$

where the definitions of λ and λ_m are same as Theorem 7.1, then the consensus of MASs will be realized.

Remark 7.8 The steps to use Corollary 7.1 are given as follows:

(1) Based on the fact that parameter λ_m is the largest eigenvalue of matrix $(b(L(m) + C(m)) + I_N)^T\,(b(L(m) + C(m)) + I_N)\,(m \neq 0)$, it is determined whether the value of λ_m is less than 1 when the MASs is under DoS attacks.

(2) If $\lambda_m < 1$, the one uses (7.33) to determine the number of impulses to be exerted in the blue window

$$z_p = \left\lceil \frac{-(\lambda_A + 2l)(\tau_n^{p+1} - \tau_n^p) - \ln(\xi \lambda^{x_p} \lambda_m^{y_p})}{\ln \lambda} \right\rceil. \tag{7.35}$$

If $\lambda_m \geq 1$, the one uses (7.34) to determine the number of impulses to be exerted in the blue window

$$z_p = \left\lceil \frac{-(\lambda_A + 2l)(\tau_n^{p+1} - \tau_n^p) - \ln(\xi \lambda^{x_p})}{\ln \lambda} \right\rceil. \tag{7.36}$$

(3) After determining the number of impulses to be applied in the blue window, it is necessary to apply an appropriate number of impulses in the blue window. Note that the instant of applying impulse does not need to be fixed, which can be exerted arbitrarily in the blue window.

Remark 7.9 Corollary 7.1 is to ensure that all exerted impulses can promote the consensus of MASs. It is necessary to ensure that for all impulsive instant, $\lambda < 1$ or $\lambda_m < 1$. In this case, from (7.13) and (7.14), one knows that after experiencing an impulsive instant, $V(t)$ must decrease, which means that the global consensus error of the MASs will also decrease. In addition, compared (7.32) to (7.36), one gets when $\lambda_m \geq 1$,

$$\begin{aligned} &\left\lceil \frac{-(\lambda_A + 2l)(\tau_n^{p+1} - \tau_n^p) - \ln(\xi \lambda^{x_p})}{\ln \lambda} \right\rceil \\ \leq &\left\lceil \frac{-(\lambda_A + 2l)(\tau_n^{p+1} - \tau_n^p) - \ln(\xi \lambda^{x_p} \lambda_m^{y_p})}{\ln \lambda} \right\rceil. \end{aligned} \tag{7.37}$$

Compared with the control strategy designed in Theorem 7.1, it can be concluded that when the impulsive control in the attack window cannot be guaranteed to promote consensus, the number of impulses that need to be exerted in the corresponding blue window will also decrease under Corollary 7.1. Therefore, adopting Corollary 7.1 can further save control resources and avoid situations where impulses in the attack window increase the global consensus error.

Remark 7.10 The Lyapunov function in Wang and He (2019) is chosen as $V(e(t)) = \sum_{i=1}^{N} e_i^T(t) P e_i(t)$, and when $t = t_k$,

$$\begin{aligned} V(e(t_k)) &= e^T(t_k^-)\Big(\Big(I_N - c\left(L_{\sigma(t_k^-)} + D_{\sigma(t_k^-)}\right)\Big) \otimes I_N\Big)^T \\ &\quad (I_N \otimes P)\left(\left(I_N - c\left(L_{\sigma(t_k^-)} + D_{\sigma(t_k^-)}\right)\right) \otimes I_N\right) e(t_k^-) \\ &\leq \sigma_{\max}^2\left(I_N - c\left(L_{\sigma(t_k^-)} + D_{\sigma(t_k^-)}\right)\right) \\ &= \mu(t_k^-) V\left(e(t_k^-)\right), \end{aligned} \tag{7.38}$$

where $\sigma(t_k^-) \in \{2, \ldots, m\}$ indicates the topology of MAS after DoS attacks. This is similar to the processing of the impulsive instant in this section, and the jump of the Lyapunov function at the impulsive instant is related to the current topology under the DoS attack. The $\mu(t_k^-)$ in the above formula is equivalent to λ_m in this section. However, in the proof of Theorem 1 in Wang and He (2019), $\mu_2 = \max \sigma_{\max}^2(I_N - c(L_{\sigma(t_k^-)} + D_{\sigma(t_k^-)}))$ is used to replace $\mu(t_k^-)$. It is obviously that $\mu_2 \geq \mu(t_k^-)$ and $\mu_2 \geq \lambda_m$. This leads to

$$\begin{aligned}
&\left\lceil \frac{-(\lambda_A + 2l)(\tau_n^{p+1} - \tau_n^p) - \ln(\xi \lambda^{x_p} \lambda_m^{y_p})}{\ln \lambda} \right\rceil \\
\leq &\left\lceil \frac{-(\lambda_A + 2l)(\tau_n^{p+1} - \tau_n^p) - \ln(\xi \lambda^{x_p} {\mu_2}^{y_p})}{\ln \lambda} \right\rceil .
\end{aligned} \tag{7.39}$$

Therefore, according to the control strategy in this subsection, the number of impulses needed in the blue window is less, so the conservativeness is lower than that in Wang and He (2019).

Remark 7.11 This section considers that DoS attacks can cause communication interruptions on some edges in MASs, rather than interrupting all communication in the entire system. This situation is somewhat similar to the switching topology problem of MASs. However, unlike the problem of switching topology, the timing of topology transformation under DoS attacks considered in this section is unpredictable, and the switched topology can be any structure that may no longer satisfy the directed spanning tree condition. Therefore, the difficulty of resisting DoS attacks investigated in this section is greater than the difficulty of the problem of switching topology.

7.3.2 *Impulsive Consensus of Multi-agent Systems Under Deception Attacks*

The value of stochastic variable $\gamma_i(t)$ is 1 or 0. The probability of $\gamma_i(t)$ is

$$\begin{aligned}
Prob\{\gamma_i(t) = 1\} &= \lambda_i, \\
Prob\{\gamma_i(t) = 0\} &= 1 - \lambda_i,
\end{aligned} \tag{7.40}$$

where $\lambda_i \in [0, 1]$.

Remark 7.12 In practice, the energy of deception attacks is mostly limited, i.e., $\|Q(t)\|^2 \leq q \ (q > 0)$.

Assumption 7.1 The stochastic variables $\gamma_i(t)$ are independent.

Denote $\Upsilon(t_k) = \mathrm{diag}(\gamma_1(t), \gamma_2(t), \ldots, \gamma_N(t))$, and $E(\Upsilon(t_k)) = \Pi$, where $\Pi = \mathrm{diag}(\lambda_1, \lambda_2, \ldots, \lambda_N)$.

The objective of the mean-squared bounded consensus control is to design distributed impulsive protocols for all followers under deception attacks, that is, for any initial conditions, the global consensus error converges to

$$\mathcal{C} = \left\{\delta(t) \in \mathbb{R}^{nN} | E\left\{\|\delta(t)\|^2\right\} < \rho\right\} \tag{7.41}$$

when $t \to \infty$, where $\delta(t) = [\delta_1^T(t), \cdots, \delta_N^T(t)]^T$, and $\rho > 0$ is error bound.

Next, this chapter will consider injection attacks and replacement attacks, respectively, for controller design and stability analysis.

(1) Controller design and stability analysis under injection attacks

Under injection attacks, the control protocol is

$$\begin{aligned} u_i(t) = \sum_{k=1}^{\infty} b \Bigg(\sum_{j \in N_i} & \left(a_{ij}\left(x_i(t) - x_j(t)\right)\right. \\ & \left. + \gamma_{ij}(t) q_i(t)\right) + c_i\left(x_i(t) - x_0(t)\right) \Bigg) \delta^*(t - t_k), \end{aligned} \tag{7.42}$$

where $k \in \mathbb{N}^+$, b is impulsive control gain. $\delta^*(t) = 0$ when $t \neq 0$, $\int_{-\infty}^{+\infty} \delta^*(t)\,dt = 1$. Assume $x_i(t_k) = x_i\left(t_k^-\right)$. Impulsive sequence $\{t_k\}_{k=1}^{\infty}$ satisfies $h_1 < t_k - t_{k-1} < h_2$, where h_1 and h_2 are positive constants.

From (7.1), (7.2), and (7.44), one gets

$$\begin{cases} \dot{\delta}_i(t) = A\delta_i(t) + f(x_i(t)) - f(x_0(t)), t \neq t_k, \\ \Delta\delta_i(t_k) = \delta_i(t_k^+) - \delta_i(t_k^-) \\ \qquad\qquad = be_i(t_k) + b\gamma_i(t_k)q_i(t_k), \end{cases} \tag{7.43}$$

where $i = 1, 2, \ldots, N$.

By using the Kronecker product, the system (7.43) can be rewritten as

$$\begin{cases} \dot{\delta}(t) = (I_N \otimes A)\delta(t) + \bar{F}(x(t), \bar{x}_0(t)), t \neq t_k, \\ \Delta\delta(t_k) = b(\Lambda \otimes I_n)\delta(t_k) + b(\Upsilon(t_k) \otimes I_n)Q(t_k), \end{cases} \tag{7.44}$$

where $x(t) = [x_1^T(t), x_2^T(t), \ldots, x_N^T(t)]^T$, $e(t) = [e_1^T(t), e_2^T(t), \ldots, e_N^T(t)]^T$, $\bar{x}_0(t) = 1_N \otimes x_0(t)$, $\bar{F}(x(t), \bar{x}_0(t)) = [f^T(x_1(t)) - f^T(x_0(t)), \ldots, f^T(x_N(t)) - f^T(x_0(t))]^T$, $\Lambda = L + C$.

The following Theorem 7.2 will give sufficient conditions for achieving mean-squared bounded consensus of MASs under injection attacks.

Theorem 7.2 *Under control input (7.42) and Assumptions 1.1, 1.2, and 7.1, if*

$$\frac{\ln \beta}{h_2} + \alpha \leq -r, \tag{7.45}$$

where $\alpha = \lambda_A + 2l > 0$, $\beta = \lambda + \varepsilon_1$, $r > 0$, λ_A *is the maximum eigenvalue of* $A + A^T$, λ *is the maximum eigenvalue of* $(I_N + b\Lambda)^T(I_N + b\Lambda)$, *the constant* $\varepsilon_1 > 0$, *then the error system (7.41) exponentially converges to* $\mathcal{C}_1 = \{\delta(t) \in \mathbb{R}^{nN} \,|\, E\{\|\delta(t)\|^2\} < \mu_1 \exp(\alpha h_2)/(1 - \beta \exp(\alpha h_2))\}$.

Proof Choose the Lyapunov function as $V(t) = \delta^T(t)\delta(t)$, when $t \in (t_{k-1}, t_k]$, one has

$$
\begin{aligned}
D^+V(t) &= \dot{\delta}^T(t)\delta(t) + \delta^T(t)\dot{\delta}(t) \\
&= \delta^T(t)(I_N \otimes (A + A^T))\delta(t) + 2\delta^T(t)\bar{f}(x(t), \bar{x}_0(t)) \\
&\leq (\lambda_A + 2l)\delta^T(t)\delta(t) \\
&= (\lambda_A + 2l)V(t),
\end{aligned}
\tag{7.46}
$$

which leads to

$$
V(t) \leq V(t_{k-1}^+)\exp((\lambda_A + 2l)(t - t_{k-1})). \tag{7.47}
$$

Take the mathematical expectation on both sides of (7.47),

$$
E(V(t)) \leq E(V(t_{k+1}^+))\exp(\alpha(t - t_k)). \tag{7.48}
$$

Denote $M = (I_N + b\Lambda) \otimes I_n$, $H(t_k) = b(\Upsilon(t_k) \otimes I_n)Q(t_k)$, then $\delta(t_k^+) = M\delta(t_k) + H(t_k)$, when $t = t_k$,

$$
\begin{aligned}
E(V(t_k^+)) &= E\left((M\delta(t_k) + H(t_k))^T (M\delta(t_k) + H(t_k))\right) \\
&= E\left(\delta^T(t_k)M^T M\delta(t_k) + \delta^T(t_k)M^T H(t_k)\right. \\
&\quad \left. + H^T(t_k)M^T\delta(t_k) + H^T(t_k)H(t_k)\right).
\end{aligned}
\tag{7.49}
$$

From the definition of M and the first term in (7.49),

$$
\begin{aligned}
&E\left(\delta^T(t_k)M^T M\delta(t_k)\right) \\
&= E\left(\delta^T(t_k)\left(\left((I_N + b\Lambda)^T(I_N + b\Lambda)\right) \otimes I_n\right)\delta(t_k)\right) \\
&\leq \lambda E(V(t_k)).
\end{aligned}
\tag{7.50}
$$

From the second and the third terms in (7.49),

$$
\begin{aligned}
&E\left(\delta^T(t_k)M^T H(t_k) + H^T(t_k)M^T\delta(t_k)\right) \\
&\leq E\left(\varepsilon_1\delta^T(t_k)\delta(t_k) + \varepsilon_1^{-1}H^T(t_k)MM^T H(t_k)\right) \\
&= E\left(\varepsilon_1\delta^T(t_k)\delta(t_k)\right) + \varepsilon_1^{-1}E\left((b\left(\Upsilon(t_k) \otimes I_n\right)Q(t_k))^T\right. \\
&\quad \left.((I_N + b\Lambda) \otimes I_n)((I_N + b\Lambda) \otimes I_n)^T b\left(\Upsilon(t_k) \otimes I_n\right)Q(t_k)\right) \\
&= \varepsilon_1 E\left(\delta^T(t_k)\delta(t_k)\right) + \varepsilon_1^{-1}E\left(\left(b^2 Q^T(t_k)\left(\Upsilon(t_k) \otimes I_n\right)\right)^T\right. \\
&\quad \left.((I_N + b\Lambda) \otimes I_n)((I_N + b\Lambda) \otimes I_n)^T\left(\Upsilon(t_k) \otimes I_n\right)Q(t_k)\right) \\
&\leq \varepsilon_1 E(V(t_k)) + \varepsilon_1^{-1}b^2 q\lambda_1,
\end{aligned}
\tag{7.51}
$$

where λ_1 is the maximum eigenvalue of $\Pi^T(I_N+b\Lambda)(I_N+b\Lambda)^T\Pi$.

From the fourth term in (7.49)

$$\begin{aligned} &E\left(H^T(t_k)H(t_k)\right)\\ &=E\left((b\left(\Upsilon(t_k)\otimes I_n\right)Q(t_k))^T\left(b\left(\Upsilon(t_k)\otimes I_n\right)Q(t_k)\right)\right)\\ &=b^2E\left(Q(t_k)^T\left(\Upsilon^T(t_k)\Upsilon(t_k)\otimes I_n\right)Q(t_k)\right)\\ &\le b^2q\lambda_{\max}\left(\Pi^T\Pi\right). \end{aligned} \tag{7.52}$$

Denote $\mu_1=\varepsilon_1^{-1}b^2q\lambda_1+b^2q\lambda_{\max}\left(\Pi^T\Pi\right)$, then

$$E(V(t_k^+))\le\beta E(V(t_k^+))+\mu_1. \tag{7.53}$$

From (7.48) and (7.53),

$$\begin{cases} E(V(t))\le E(V(t_k^+))\exp(\alpha(t_k-t_{k-1})),\ t\neq t_k,\\ E(V(t_k^+))\le\beta E(V(t_k))+\mu_1. \end{cases} \tag{7.54}$$

When $t\in(t_0,t_1]$,

$$E(V(t))\le E(V(t_0))\exp(\alpha(t-t_0)). \tag{7.55}$$

When $t\in(t_1,t_2]$,

$$\begin{aligned} E(V(t))&\le E(V(t_1^+)\exp(\alpha(t-t_1))\\ &\le\beta E(V(t_1))\exp(\alpha(t-t_1))+\mu_1\exp(\alpha(t-t_1))\\ &\le\beta E(V(t_0))\exp(\alpha(t-t_0))+\mu_1\exp(\alpha(t-t_1)). \end{aligned} \tag{7.56}$$

When $t\in(t_2,t_3]$,

$$\begin{aligned} E(V(t))&\le E(V(t_2^+))\exp(\alpha(t-t_2))\\ &\le\beta E(V(t_2))\exp(\alpha(t-t_2))+\mu_1\exp(\alpha(t-t_2))\\ &\le\beta^2E(V(t_0))\exp(\alpha(t-t_0))+\beta\mu_1\exp(\alpha(t-t_1))\\ &\quad+\exp(\alpha(t-t_2)). \end{aligned} \tag{7.57}$$

In general, when $t\in(t_k,t_{k+1}]$,

$$\begin{aligned}E(V(t)) &\leq \beta^k E(V(t_0))\exp(\alpha(t-t_0)) + \beta^{k-1}\mu_1 \exp(\alpha(t-t_1))\\&\quad + \beta^{k-2}\mu_1 \exp(\alpha(t-t_2)) + \cdots + \mu_1 \exp(\alpha(t-t_k))\\&\leq \beta^k E(V(t_0))\exp(\alpha(t-t_0)) + \mu_1 \left(\beta^{k-1}\exp(\alpha(t_{k+1}-t_1))\right.\\&\quad \left.+\beta^{k-2}\exp(\alpha(t_{k+1}-t_2)) + \cdots + \exp(\alpha(t_{k+1}-t_k))\right)\\&\leq \beta^k E(V(t_0))\exp(\alpha(t-t_0)) + \mu_1 \left(\beta^{k-1}\exp(\alpha k h_2)\right.\\&\quad \left.+\beta^{k-1}\exp(\alpha(k-1)h_2) + \cdots + \exp(\alpha h_2)\right)\\&= \beta^k E(V(t_0))\exp(\alpha(t-t_0))\\&\quad + \mu_1 \frac{\exp(\alpha h_2) - \exp(\alpha h_2)\exp((\alpha h_2 + \ln\beta)k)}{1-\beta\exp(\alpha h_2)}.\end{aligned} \tag{7.58}$$

Since $k > \frac{t-t_0}{h_2} - 1$, from (7.58),

$$\begin{aligned}E(V(t)) &\leq \beta^{\frac{t-t_0}{h_2}-1} E(V(t_0))\exp(\alpha(t-t_0)) + \mu_1 \frac{\exp(\alpha h_2)}{1-\beta\exp(\alpha h_2)}\\&\quad - \mu_1 \frac{\exp(\alpha h_2)\exp\left((\alpha h_2 + ln\beta)\left(\frac{t-t_0}{h_2}-1\right)\right)}{1-\beta\exp(\alpha h_2)}\\&= \exp\left(\left(\frac{t-t_0}{h_2}-1\right) ln\beta + \alpha(t-t_0)\right) E(V(t_0))\\&\quad + \mu_1 \frac{\exp(\alpha h_2)}{1-\beta\exp(\alpha h_2)} - \mu_1 \frac{\exp\left(\left(\frac{\ln\beta}{h_2}+\alpha\right)(t-t_0) - \ln\beta\right)}{1-\beta\exp(\alpha h_2)}.\end{aligned} \tag{7.59}$$

From (7.45), one gets

$$\begin{aligned}E(V(t)) \leq \frac{\exp(-r(t-t_0))}{\beta} E(V(t_0)) + \frac{\mu_1 \exp(\alpha h_2)}{1-\beta\exp(\alpha h_2)}\\- \mu_1 \exp(-r(t-t_0)) \frac{-\ln\beta}{1-\exp(\ln\beta + \alpha h_2)}.\end{aligned} \tag{7.60}$$

Therefore, when $t \to \infty$, the error system (7.41) converges to the ball

$$\mathcal{C}_1 = \left\{ E\left\{\|\delta(t)\|^2\right\} < \frac{\mu_1 \exp(\alpha h_2)}{1-\beta\exp(\alpha h_2)} \right\}. \tag{7.61}$$

The proof of Theorem 7.2 is completed.

Remark 7.13 In Wen et al. (2020), the authors obtained

$$E(V(t)) \leq E(V(t_k^+))\exp(-\mu(t-t_{k-1})), \tag{7.62}$$

for $t \in [t_{k-1}, t_k)$, where μ is a positive constant. From (7.62), one obtains $\lim_{t\to\infty} E(V(t)) = 0$, then the MASs realizes mean-squared consensus asymptotically without any control input. However, it is difficult to achieve consensus of MASs without control input in practice. Therefore, $\eta > 0$ is considered in this subsection.

Remark 7.14 From (7.45), the upper bound of impulsive control interval is

$$h_2 \leq -\frac{\ln \beta}{\alpha + r}. \tag{7.63}$$

From (7.60), r represents the convergence rate. From (7.63), when r increases, the upper bound of impulsive interval h_2 decreases, which presents that if one wants to get a faster convergence rate, the impulsive interval should be chosen smaller. From (7.61), one knows h_2 is related to the error bound ρ_1. If h_2 becomes smaller, then the error bound $\rho_1 = \mu_1 \exp(\alpha h_2)/(1 - \beta \exp(\alpha h_2))$ becomes smaller. That's to say, if r increases, the error bound ρ_1 decreases. Therefore, the selection of parameter r is a trade-off between obtaining a smaller error bound and obtaining larger impulsive intervals. In short, to use Theorem 7.1, we should select the appropriate convergence rate r and impulsive control gain b firstly, and calculate the values of α and β, then h_2 is obtained by (7.63).

Remark 7.15 To guarantee the mean-squared bounded consensus is realized, $\beta \in (0, 1)$ is a necessary condition. From (7.63), one gets if $\beta \geq 1$, then $-\frac{\ln \beta}{\alpha + r} \leq 0$. In practice, the impulsive interval $t_k - t_{k-1}$ cannot be zero or less than zero. Therefore, we should select an appropriate impulsive control gain b to make sure $\beta \in (0, 1)$. In addition, if the value of parameter β becomes smaller, then after an impulsive control, the global synchronization error of MASs becomes smaller. Therefore, to achieve a better control performance, we can choose impulsive control gain b to minimize the value of β.

Remark 7.16 From (7.61), one knows h_1 is not related to error bound ρ_1. If one chooses a smaller h_1, then the impulsive intervals become smaller, and the conservatism is increased. Therefore, to reduce the conservatism, we choose $h_1 = h_2 = -\frac{\ln \beta}{\alpha + r}$ in this subsection.

(2) Controller design and stability analysis under replacement attacks

Under the replacement deception attacks, the control protocol of agent i is

$$u_i(t) = \sum_{k=1}^{\infty} (\gamma_i(t) q_i(t) + (1 - \gamma_i(t)) b e_i(t_k)) \, \delta^*(t - t_k), \tag{7.64}$$

where $i = 1, 2, \ldots, N$.

Rewritten (7.64), one gets

$$u_i(t) = \begin{cases} \sum\limits_{k=1}^{\infty} b e_i(t_k) \delta^*(t - t_k), \ \gamma_i(t) = 0, \\ \sum\limits_{k=1}^{\infty} \gamma_i(t) q_i(t) \delta^*(t - t_k), \ \gamma_i(t) = 1. \end{cases} \tag{7.65}$$

From (7.65), one knows if $\gamma_i(t) = 1$, the control input of agent i is completely replaced with the attack signal $\gamma_i(t)q_i(t)$.

From the system model (7.1), (7.2), and the control input (7.64),

$$\begin{cases} \dot{\delta}_i(t) = A\delta_i(t) + f(x_i(t)) - f(x_0(t)),\ t \neq t_k, \\ \Delta\delta_i(t_k) = \delta_i(t_k^+) - \delta_i(t_k^-) \\ \qquad\qquad = \gamma_i(t_k)q_i(t_k) + b(1-\gamma_i(t_k))e_i(t_k), \end{cases} \tag{7.66}$$

where $i = 1, 2, \ldots, N$.

By using the Kronecker product, the system (7.66) can be rewritten as

$$\begin{cases} \dot{\delta}(t) = (I_N \otimes A)\delta(t) + \bar{F}(x(t), \bar{x}_0(t)),\ t \neq t_k, \\ \Delta\delta(t_k) = (b\Lambda - \bar{\Upsilon}(t_k)b\Lambda \otimes I_n)\delta(t_k) + (\bar{\Upsilon}(t_k) \otimes I_n)Q(t_k). \end{cases} \tag{7.67}$$

In order to obtain sufficient conditions for achieving mean-squared bounded consensus of MASs under replacement attacks, the following Theorem 7.3 below will be given.

Theorem 7.3 *Under control input (7.64) and Assumptions 1.1, 1.2, and 7.1, if*

$$\frac{\ln \bar{\beta}}{h_2} + \alpha \leq -r, \tag{7.68}$$

where the definitions of α *and* r *are same as Theorem 7.2,* $\bar{\beta} = \bar{\lambda} + \varepsilon_2$, $\bar{\lambda}$ *is the maximum eigenvalue of* $(I_N + b\Lambda - b\Pi\Lambda)^T(I_N + b\Lambda - b\Pi\Lambda)$, *the constant* $\varepsilon_2 > 0$, *then system (7.67) exponentially converges to* $\mathcal{C}_2 = \{\delta(t) \in \mathbb{R}^{nN} \,|\, E\{\|\delta(t)\|^2\} < \mu_2 \exp(\alpha h_2)/(1 - \beta\exp(\alpha h_2))\}$, *the definition of* μ_2 *will be given in the proof of Theorem 7.3.*

Proof Choose the Lyapunov function as $V(t) = \delta^T(t)\delta(t)$, similar to the proof of Theorem 7.2, when $t \in (t_{k-1}, t_k]$,

$$E(V(t)) \leq E(V(t_{k+1}^+))\exp(\alpha(t - t_k)). \tag{7.69}$$

When $t = t_k$,

$$\begin{aligned} \delta(t_k^+) &= \delta(t_k) + ((b\Lambda - b\Upsilon(t_k)\Lambda) \otimes I_n)\,\delta(t_k) + (\Upsilon(t_k) \otimes I_n)\,Q(t_k) \\ &= ((I_N + b\Lambda - b\Upsilon(t_k)\Lambda) \otimes I_n)\,\delta(t_k) + (\Upsilon(t_k) \otimes I_n)\,Q(t_k) \\ &= \bar{M}\delta(t_k) + H(t_k), \end{aligned} \tag{7.70}$$

where $\bar{M} = (I_N + b\Lambda - b\Upsilon(t_k)\Lambda) \otimes I_n$, the definition of $H(t_k)$ is same as Theorem 7.2, then one gets

$$
\begin{aligned}
E(V(t_k^+)) &= E\left(\left(\bar{M}\delta(t_k) + H(t_k)\right)^T \left(\bar{M}\delta(t_k) + H(t_k)\right)\right) \\
&= E\left(\delta^T(t_k)\bar{M}^T\bar{M}\delta(t_k) + \delta^T(t_k)\bar{M}^T H(t_k)\right. \\
&\quad \left. + H^T(t_k)\bar{M}^T\delta(t_k) + H^T(t_k)H(t_k)\right).
\end{aligned} \tag{7.71}
$$

For the first term in (7.71),

$$
\begin{aligned}
&E\left(\delta^T(t_k)\bar{M}^T\bar{M}\delta(t_k)\right) \\
&= E\left(\delta^T(t_k)\left(\left((I_N + b\Lambda - b\Pi\Lambda)^T(I_N + b\Lambda - b\Pi\Lambda)\right) \otimes I_n\right)\delta(t_k)\right) \\
&\leq \bar{\lambda}E(V(t_k)).
\end{aligned} \tag{7.72}
$$

For the second and third terms in (7.72),

$$
\begin{aligned}
&E\left(\delta^T(t_k)\bar{M}^T H(t_k) + H^T(t_k)\bar{M}^T\delta(t_k)\right) \\
&\leq E\left(\varepsilon_2\delta^T(t_k)\delta(t_k) + \varepsilon_2^{-1}H^T(t_k)\bar{M}\bar{M}^T H(t_k)\right) \\
&= \varepsilon_2^{-1}E\left(\left((\Upsilon(t_k) \otimes I_n)\, Q(t_k)\right)^T ((I_N + b\Lambda - b\Upsilon(t_k)\Lambda) \otimes I_n)\right. \\
&\quad \left. ((I_N + b\Lambda - b\Upsilon(t_k)\Lambda) \otimes I_n)^T (\Upsilon(t_k) \otimes I_n)\, Q(t_k)\right) + E\left(\varepsilon_2\delta^T(t_k)\delta(t_k)\right) \\
&= \varepsilon_2^{-1}E\left(\left(Q^T(t_k)\,(\Upsilon(t_k) \otimes I_n)\right)^T ((I_N + b\Lambda - b\Upsilon(t_k)\Lambda) \otimes I_n)\right. \\
&\quad \left. ((I_N + b\Lambda - b\Upsilon(t_k)\Lambda) \otimes I_n)^T (\Upsilon(t_k) \otimes I_n)\, Q(t_k)\right) + E\left(\varepsilon_2\delta^T(t_k)\delta(t_k)\right) \\
&\leq \varepsilon_2 E(V(t_k)) + \varepsilon_2^{-1}q\lambda_2,
\end{aligned} \tag{7.73}
$$

where the parameter λ_2 is the maximum eigenvalue of the matrix $\Pi^T(I_N + b\Lambda - b\Pi\Lambda)(I_N + b\Lambda - b\Pi\Lambda)^T\Pi$.

For the fourth term in (7.73)

$$
\begin{aligned}
E\left(H^T(t_k)H(t_k)\right) &= E\left(\left((\Upsilon(t_k) \otimes I_n)\, Q(t_k)\right)^T \left((\Upsilon(t_k) \otimes I_n)\, Q(t_k)\right)\right) \\
&= E\left(Q(t_k)^T\left(\Upsilon^T(t_k)\Upsilon(t_k) \otimes I_n\right) Q(t_k)\right) \\
&\leq q\lambda_{\max}\left(\Pi^T\Pi\right).
\end{aligned} \tag{7.74}
$$

Denote $\mu_2 = \varepsilon_2^{-1}q\lambda_2 + q\lambda_{\max}\left(\Pi^T\Pi\right)$, then one gets

$$
E(V(t_k^+)) = \bar{\beta}E(V(t_k)) + \mu_2. \tag{7.75}
$$

From (7.69) and (7.75), one gets

$$
\begin{cases}
E(V(t)) \leq E(V(t_k^+))\exp(\alpha(t_k - t_{k-1})),\, t \neq t_k, \\
E(V(t_k^+)) \leq \bar{\beta}E(V(t_k)) + \mu_2.
\end{cases} \tag{7.76}
$$

The rest recursive analysis of the proof is similar to Theorem 7.2. The proof is completed.

Remark 7.17 In He et al. (2020), the stochastic variables $\beta_i(t)$ (the variable $\gamma_i(t)$ in this subsection) are represented as

$$\begin{aligned} Prob\{\beta_i(t) = 1\} &= \bar{\beta}, \\ Prob\{\beta_i(t) = 0\} &= 1 - \bar{\beta}, \end{aligned} \tag{7.77}$$

where $\bar{\beta} \in [0, 1]$ is a constant. In He et al. (2020), the attack probability of all agents in MASs is the same. In practice, because each agent in MASs is in a different environment, the probability of suffering deception attacks should also be different. Compared with He et al. (2020), in this subsection, the attack probability of each agent in MASs is different, which is more reasonable in practice.

Remark 7.18 Comparing the results obtained in this section with those in previous chapters and under DoS attacks, it can be seen that when the MASs are subjected to deception attacks, applying impulsive control can only achieve mean-squared bounded consensus of the MASs and cannot achieve asymptotic consensus. This also indicates that in the problems of impulsive consensus of MASs, the situation where the MASs are subjected to deception attacks is more severe than that of DoS attacks, and it is more difficult to eliminate the impact of deception attacks on the stability of error systems. In practice, compared to asymptotic consensus, the problem of mean-squared bounded consensus is more common when a system is under attack or disturbance. However, if the attack signal is very strong, i.e., the value of q is very large, then as the definitions of constants μ_1 and μ_2, the values of μ_1 and μ_2 will also become very large. From Theorems 7.2 and 7.3, it can be concluded that when the values of μ_1 and μ_2 become very large, the value of the error bound ρ will also become very large, even causing the obtained results to lose practical significance. According to the analysis in Remark 7.14, when the error bound is large, it can be reduced by increasing the convergence rate r of the system. However, the increase of convergence rate r will decrease the calculated upper bound of impulsive interval, which means that the consumption of impulsive control input needs to be increased. This is the limitation of the impulsive control protocol designed in this subsection to deal with deception attacks.

7.4 Simulation Results

(1) Simulation for impulsive consensus control under DoS attacks

Assuming the MASs contain four followers and one leader, the system model used in this section is the same as in Chap. 2 and still follows Chua's circuit as follows:

$$\begin{cases} \dot{x}_{i1} = p_1(x_{i2} - x_{i1} - g(x_{i1})), \\ \dot{x}_{i2} = x_{i1} - x_{i2} + x_{i3}, \\ \dot{x}_{i3} = -p_{i2}x_{i2}, \end{cases} \tag{7.78}$$

where p_1 and p_2 are system parameters, the expression of nonlinear function $g(\cdot)$ is

$$g(x_{i1}) = m_2 x_{i1} + 0.5(m_1 - m_2)\left(|x_{i1} + 1| - |x_{i1} - 1|\right), \tag{7.79}$$

where $m_1 < m_2 < 0$ are two given constants.

Consider the topology of MASs in Fig. 7.2. There are six edges existing among these nodes for information transmission, which are represented as ed1, ed2, ed3, ed4, ed5, and ed6. Here, there are four situations that communication channels (edges) are attacked. To be specific, in Fig. 7.3(a), the edges ed3 and ed5 are attacked, and the information transmission from the agent 1 to 2 and from the agent 2 to 4 are interrupted. Thus, the update of nodes 2 and 4 are affected by DoS attacks, but the other agents are not. In Fig. 7.3(d), all edges are attacked by DoS attacks, and the communication between nodes is completely interrupted.

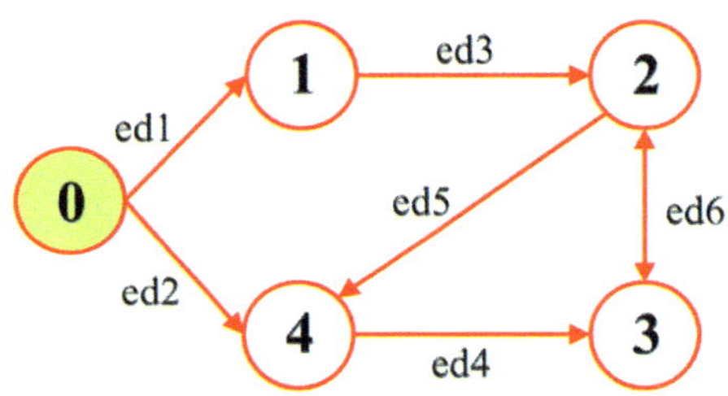

Fig. 7.2 Topology graph of the multi-agent systems in Example 7.1

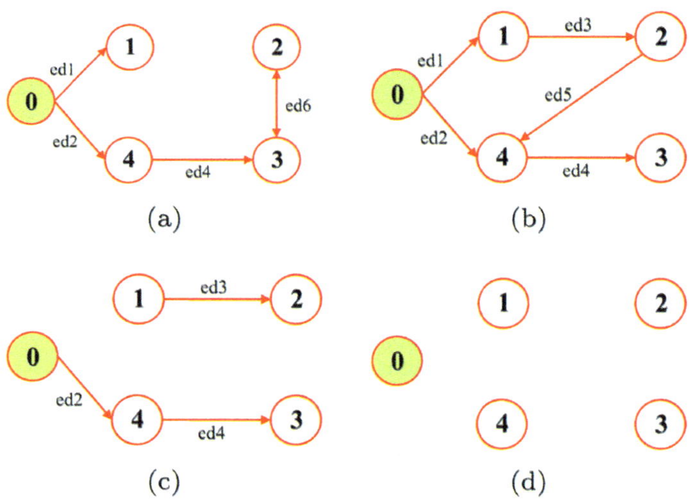

Fig. 7.3 Different edges are attacked by DoS attacks

Example 7.1 (*Adopt the control strategy designed in Theorem* 7.1) From (7.13), in order to achieve the consensus of MASs, the selection of impulsive control gain must satisfy $\lambda \in (0, 1)$, because only when $\lambda \in (0, 1)$, the impulsive control applied in the green and blue windows is guaranteed to help achieve the consensus of MASs. The allowable range of impulsive control gain b is presented in Fig. 7.4. From Fig. 7.4, one gets the range of b is $b \in (-0.5787, 0)$, and when $b = -0.5$, the value of λ is smallest. From (7.10) in Theorem 7.1, one gets if λ decreases, $\bar{\lambda}_p$ decreases, then the length of window $T_p - T_{p-1}$ increases, which means that the allowable range of impulsive instant is larger. Therefore, we choose $b = -0.5$ in this section.

Let $\xi = 1.01$, from (7.10) in Theorem 7.1, one obtains the number of impulses in blue window is

$$z_p \geq \frac{-(\lambda_A + 2l)(\tau_n^{p+1} - \tau_n^p) - ln(1.01\lambda^{x_p}\lambda_m^{y_p})}{\ln \lambda}.$$

From the discussion in Remark 7.2, choose $z_p = \left\lceil \frac{-(\lambda_A+2l)(\tau_n^{p+1}-\tau_n^p)-ln(\xi\lambda^{x_p}\lambda_m^{y_p})}{\ln \lambda} \right\rceil$, then the impulsive interval in blue window can be set as $(\tau_n^{p+1} - \tau_n^p) / z_p$.

In this section, it is assumed that the minimum interval between two adjacent attacks is 0.05 s. Figure 7.5 shows that under the distributed impulsive control law designed in this chapter, all following agents can achieve consensus with the leader. The curve of the obtained impulsive interval is shown in Fig. 7.6.

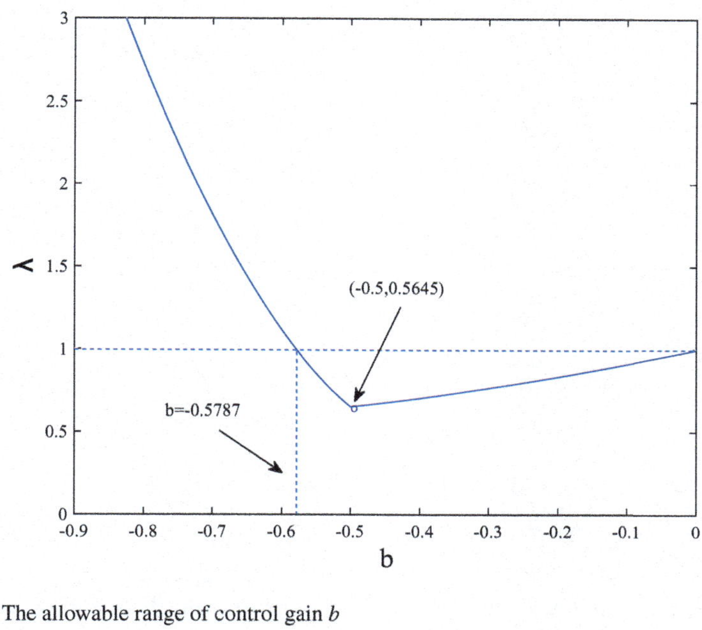

Fig. 7.4 The allowable range of control gain b

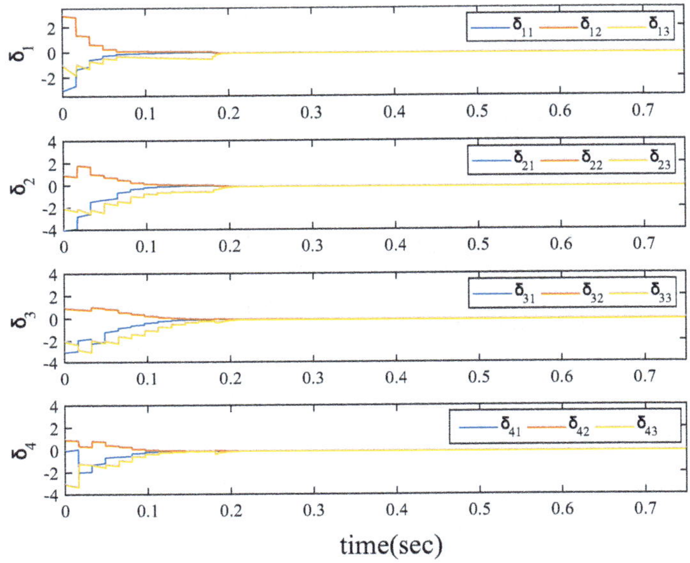

Fig. 7.5 Consensus error for the result in Theorem 7.1

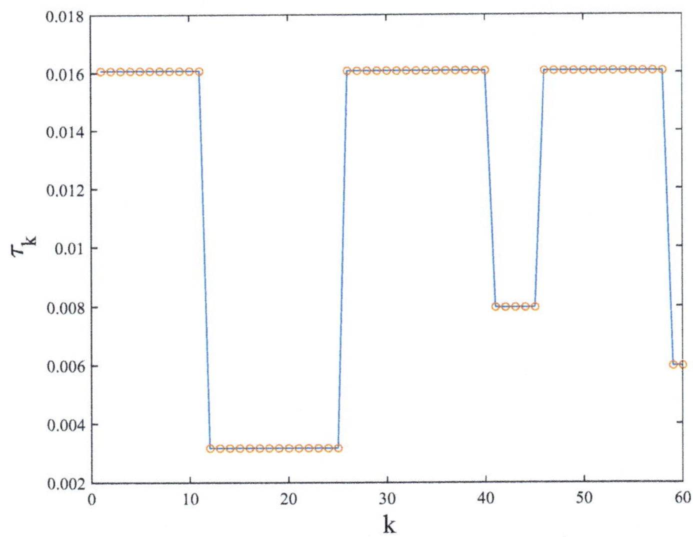

Fig. 7.6 The impulsive interval $\tau_k = t_{k+1} - t_k$ versus k in Theorem 7.1

Figure 7.7 is the schematic diagram of the DoS attacks in Example 7.1, which presents the time, duration, and topology under the DoS attack. In Fig. 7.7, the green,

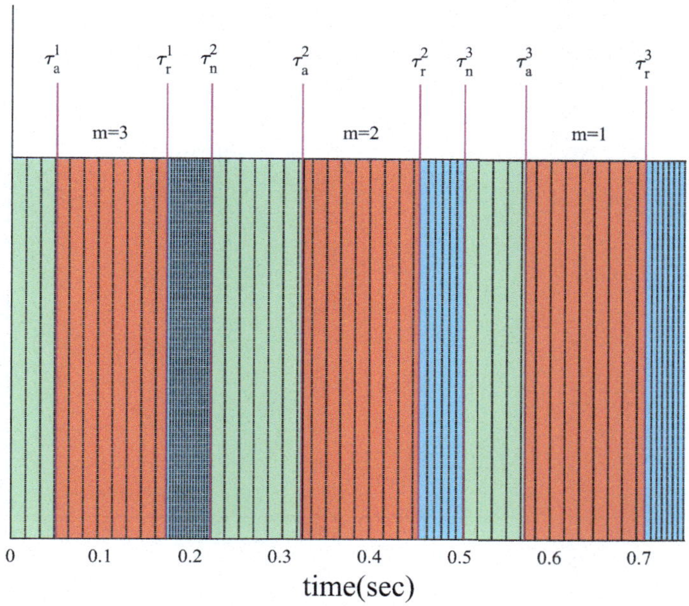

Fig. 7.7 The diagram of DoS attack of Theorem 7.1

red, and blue areas represent the normal window, attack window, and window, respectively. The dashed line represents the moment when impulsive control occurs. $m = 0$ indicates that the MASs have not been attacked by DoS. $m = 1, m = 2, m = 3$, and $m = 4$ correspond to (a), (b), (c), and (d) in Fig. 7.3, respectively, representing the topology of the MASs under different DoS attacks.

Example 7.2 (*Adopt the control strategy designed in* Wang and He (2019) *and Corollary* 7.1) Keeping the occurrence and duration of DoS attacks, as well as the system and control parameters unchanged, if the method proposed in Wang and He (2019) is adopted, which only considers the case of maximum λ_m, the obtained impulsive interval is shown in Fig. 7.8. Comparing Fig. 7.8 with Fig. 7.6, it can be observed that when the MASs is subjected to a second DoS attack, the number of impulses blue window obtained is much larger than when considering the actual situation of each attack, and the impulsive interval obtained is much smaller than when considering the actual situation each time. As shown in Fig. 7.7, when the MASs are subjected to the second attack, $m = 2$, then one gets $\lambda_m = 0.75$. Due to the fact that Wang and He (2019) only consider the maximum case of λ_m, i.e., $\lambda_m = 1.309$, as discussed in Remark 7.9, this will result in a larger number of impulses being calculated in the blue window. That is to say, if the actual situation of each attack is considered separately, the required number of impulses in the blue window is relatively small, and the impulsive interval can also be relatively large, thereby saving the consumption of the control input.

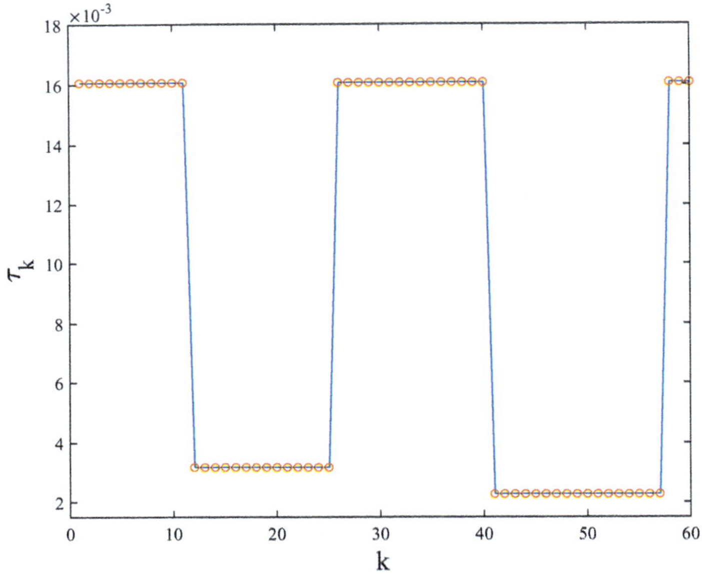

Fig. 7.8 The impulsive interval $\tau_k = t_{k+1} - t_k$ versus k in Wang and He (2019)

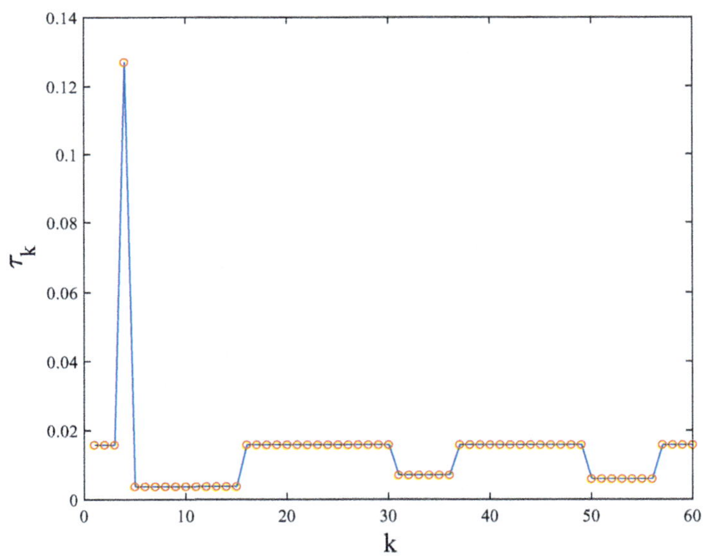

Fig. 7.9 The impulsive interval $\tau_k = t_{k+1} - t_k$ versus k in Corollary 7.1

The curve of impulsive interval by using Corollary 7.1 is presented in Fig. 7.9. From Fig. 7.7, one knows when the MASs under the first attack, $m = 3$, and then

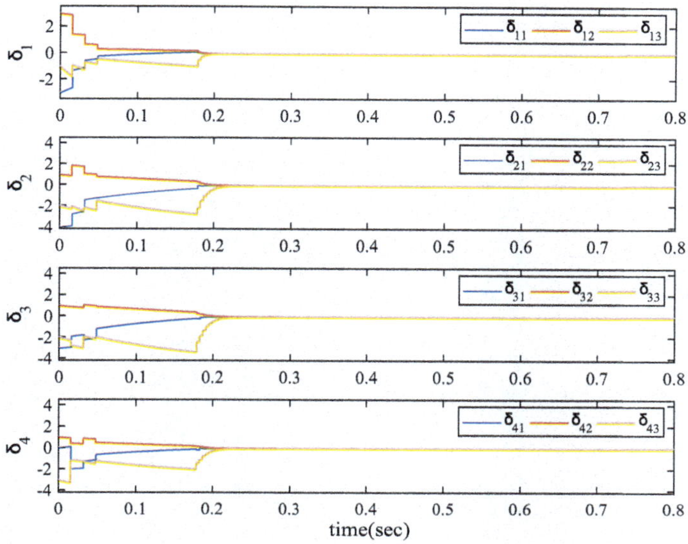

Fig. 7.10 Consensus error for the result in Corollary 7.1

$\lambda_m = 1.309 > 1$. According to Corollary 7.1, the impulsive control under the frits attack is not applied, so there is a point in Fig. 7.9 with a larger impulsive interval. After this large impulsive interval, it becomes the impulsive interval in the blue window. The impulsive interval of the first blue window in Fig. 7.6 is 0.0032 s, and the impulsive interval of the first blue window in Fig. 7.9 is 0.004 s. Since the time length of the blue window remains unchanged, this means that using Corollary 7.1, the number of impulses in the blue window is less, which is consistent with the discussion in Remark 7.8.

Figure 7.10 shows the curve of global consensus error of the MASs obtained by using Corollary 7.1. It can be seen that using the control strategy in Corollary 7.1 can also achieve consensus of the MASs under DoS attacks. Figure 7.11 is the schematic diagram of the DoS attack by using Corollary 7.1. It can be seen that when $m = 3$, $\lambda_m = 1.309 > 1$. Therefore, during the first DoS attack, there was no impulsive control input in the attack window.

Combining Figs. 7.9, 7.10, and 7.11, it can be seen that using the control scheme in Corollary 7.1, less impulses need to be applied during the control process. This not only achieves the consensus of MASs, but also saves control resources.

(2) Simulation for impulsive consensus control under deception attacks

This section will provide two simulation examples to verify the effectiveness of the distributed impulsive control algorithm designed under the two types of deception attacks mentioned above. The system models of each node considered in this section are the same as those in Examples 7.1 and 7.2.

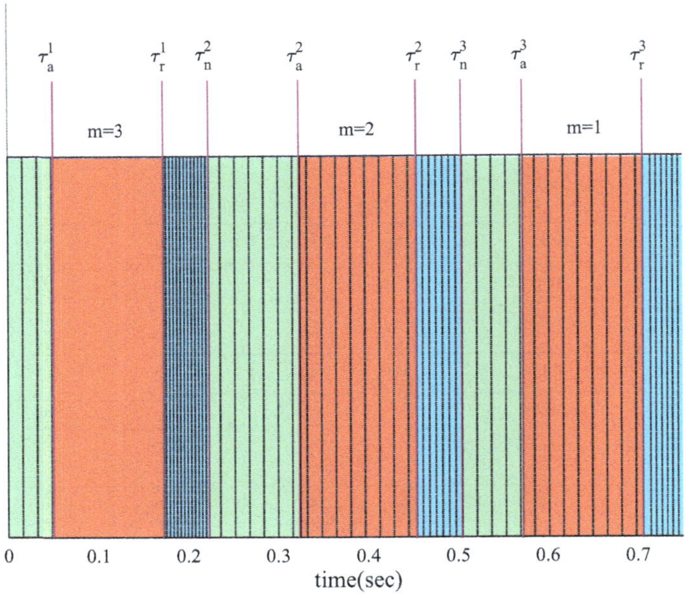

Fig. 7.11 The diagram of DoS attacks of Corollary 7.1

Fig. 7.12 Topology graph of the multi-agent systems in Example 7.2

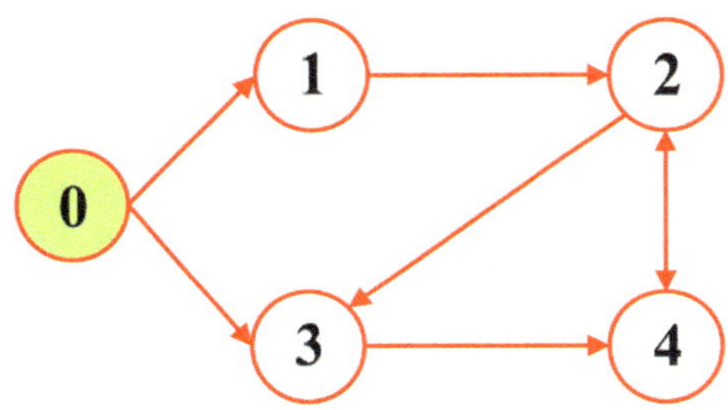

Consider the MASs with four follower nodes and one leader node, and the topology is shown in Fig. 7.12. Then one gets

$$\mathcal{A} = \begin{bmatrix} 0\,0\,0\,0 \\ 1\,0\,0\,1 \\ 0\,1\,0\,0 \\ 0\,1\,1\,0 \end{bmatrix}, \mathcal{D} = \begin{bmatrix} 0\,0\,0\,0 \\ 0\,2\,0\,0 \\ 0\,0\,1\,0 \\ 0\,0\,0\,2 \end{bmatrix},$$

$$L = \mathcal{D} - \mathcal{A} = \begin{bmatrix} 0 & 0 & 0 & 0 \\ -1 & 2 & 0 & -1 \\ 0 & -1 & 1 & 0 \\ 0 & -1 & -1 & 2 \end{bmatrix}, C = \begin{bmatrix} 1\,0\,0\,0 \\ 0\,0\,0\,0 \\ 0\,0\,1\,0 \\ 0\,0\,0\,0 \end{bmatrix}.$$

Example 7.3 (*The MASs are under inject attacks*) Choose $\varepsilon_1 = 0.1$, $q = 0.1$, and the expectation of $\Upsilon(t_k)$ is

$$E(\Upsilon(t_k)) = \Pi = \text{diag}(0.45, 0.42, 0.37, 0.52).$$

From Remark 7.15, one knows the choice of impulsive control gain b needs to guarantee $\beta \in (0, 1)$. Figure 7.13 gives the allowable range of b. From Fig. 7.13, one gets b can be selected arbitrarily in $(-0.56, -0.12)$. What's more, the β is smallest when $b = -0.5$ in Fig. 7.13. From the discussion in Remark 7.15, the impulsive control gain b is selected as $b = -0.5$ in this example, then $\beta = 0.7545$.

From the definition of α, then $\alpha = \lambda_A + 2l = 25.6118$. Choose converge rate $r = 3$, from (7.45), one gets $h_2 \leq -\ln \beta/(\alpha + r) = -\ln 0.7545/(25.6118 + 3) = 0.0098$. Based on the discussion in Remark 7.16, to reduce conservatism and minimize input loss of impulsive control, in this example, the upper and lower bounds of the impulsive interval are selected as $h_1 = h_2 = 0.0098$, i.e., choose the impulsive interval as $\tau_k = t_k - t_{k-1} = 0.0098$.

In Example 7.3, choose the initial states of each agent in MASs as $x_0 = [5.5, -5.8, 3.2]^T$, $x_1 = [3, -2.5, 3.5]^T$, $x_2 = [2.8, -1.5, 2.8]^T$, $x_3 = [8.5, -7.9, 3.7]^T$, $x_4 = [7.7, -8.2, 2.7]^T$. The curve of consensus error is shown in Fig. 7.14, and the curve of $\|\delta(t)\|^2$ is presented in Fig. 7.15. From Fig. 7.15, one gets the error bound in this example is $\rho_1 = 7.7414$ and also knows $\|\delta(t)\|^2$ can converge to the error bound. Therefore, by using the impulsive control algorithm designed

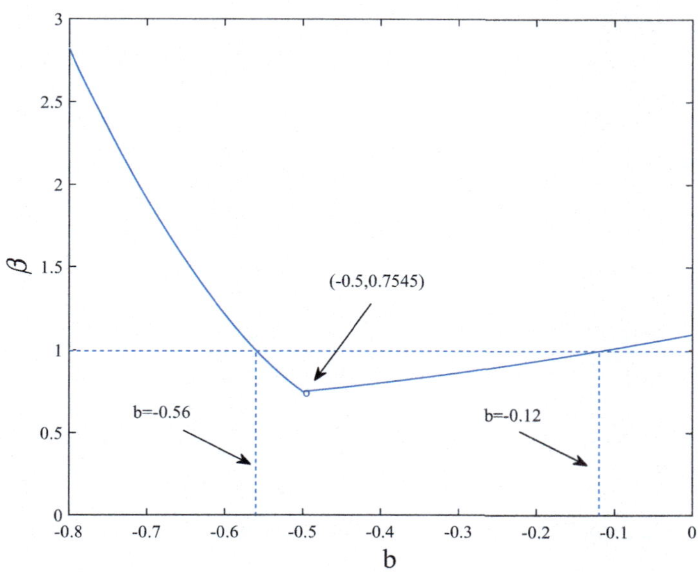

Fig. 7.13 Allowable range of impulsive control gain b in Example 7.3

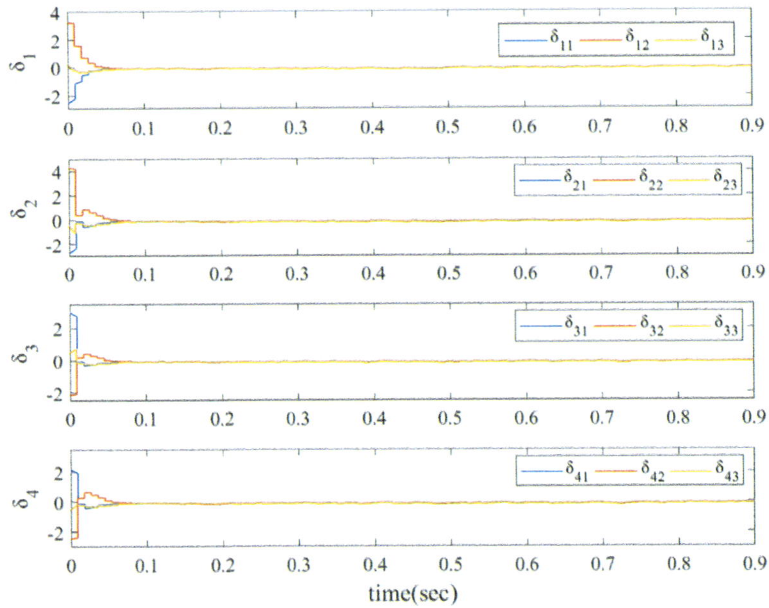

Fig. 7.14 The global consensus errors of multi-agent systems in Example 7.3

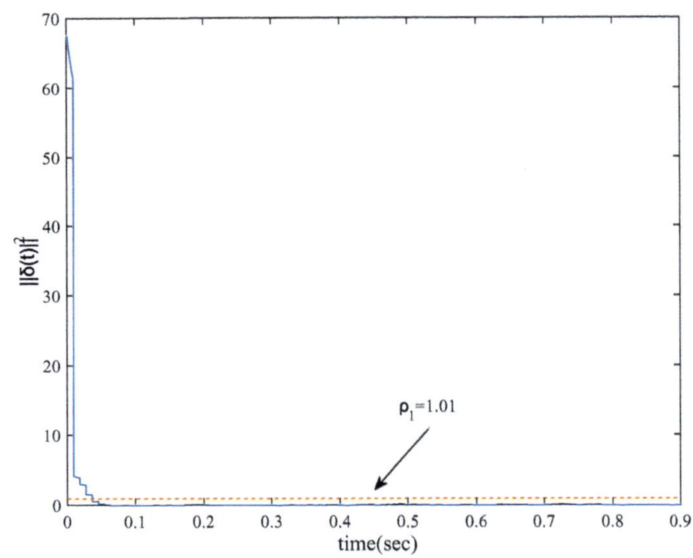

Fig. 7.15 The curves of $\|\delta(t)\|^2$ and error bound in Example 7.3

in this chapter to deal with inject attacks, the mean-squared bounded consensus of MASs can be realized, and the validity of Theorem 7.2 is verified.

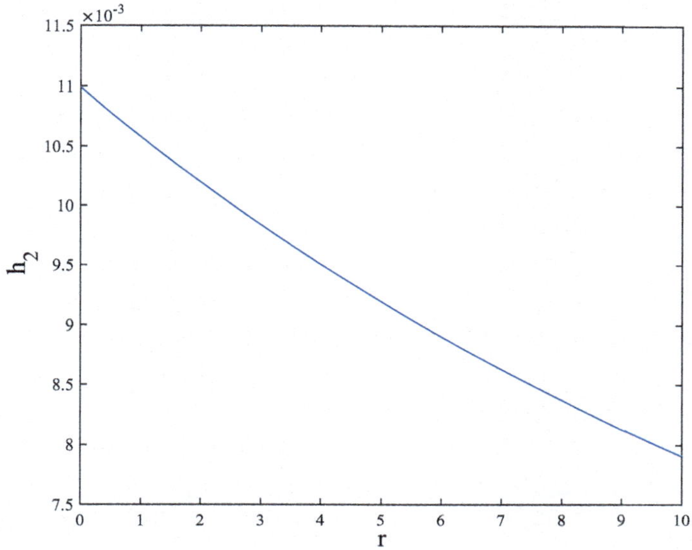

Fig. 7.16 The relationship of convergence rate r and h_2 in Example 7.3

Figure 7.16 is the relationship between convergence rate r and the upper bound of impulsive intervals h_2. From Fig. 7.16, one gets when the parameter r increases, h_2 decreases. Figure 7.17 shows the curve of r and the error bound ρ. From Fig. 7.17, one obtains if the convergence rate r increases, the error bound ρ decreases. In addition, it should be noted that as shown in Fig. 7.17, when the convergence rate r is small, the error bound ρ becomes large. When the error bound is particularly large, even if $\|\delta(t)\|^2$ can be guaranteed within the error bound, the control performance may not be very good. From Figs. 7.16 and 7.17, one gets the selection of parameter r should be a trade-off between obtaining larger impulsive intervals and obtaining a smaller error bound. Therefore, it is very important to determine the value of the exponential convergence rate r, and the parameter r is introduced in Eq. (7.45) of Theorem 7.2.

Example 7.4 (*The MASs are under replacement attacks*) Consider the same topology as Example 7.3. Choose $\varepsilon_4 = 0.1$, $q = 0.1$. The expectation of $\bar{\Upsilon}(t_k)$ is same as Example 7.3. The allowable range of impulsive control gain b in Example 7.4 is presented in Fig. 7.18. That is to say, one can choose the control gain b arbitrarily in $(-1.017, -0.217)$. It can be obtained from Fig. 7.18 that at the point $(-0.841, 0.7953)$, $\bar{\beta}$ is smallest. From the discussion in Remark 7.15, the control gain is chosen as $b = -0.841$ in Example 7.4.

Choose $r = 3$, from (7.68) in Theorem 7.3, one gets $h_2 \leq -\ln \bar{\beta}/(\alpha + r) = -\ln 0.7953/(25.6118 + 3) = 0.008$. From Remark 7.16, to reduce the conservatism, we choose $h_1 = h_2 = 0.008$.

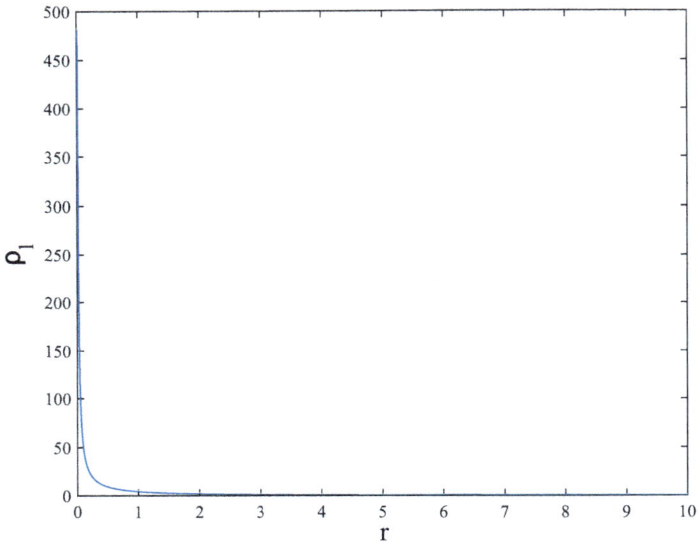

Fig. 7.17 The relationship of convergence rate r and error bound ρ_1 in Example 7.3

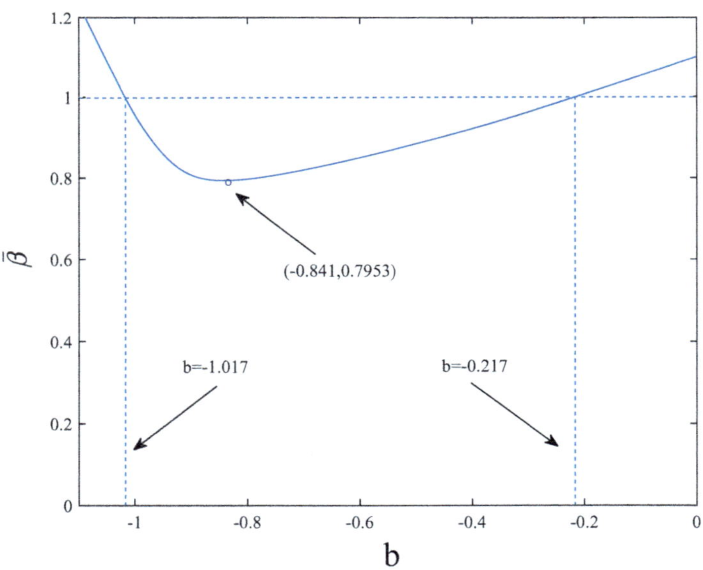

Fig. 7.18 Allowable range of impulsive control gain b in Example 7.4

In this example, the initial values are the same as Theorem 7.3. The global synchronization error in Theorem 7.3 is shown in Fig. 7.19, and the curve of variable

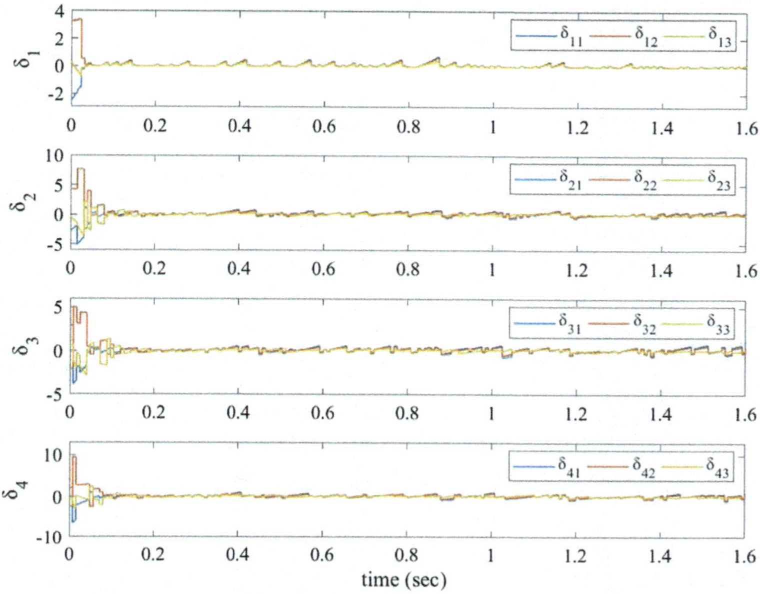

Fig. 7.19 The global consensus errors of multi-agent systems in Example 7.4

$\|\delta(t)\|^2$ is presented in Fig. 7.20, which verifies that $\|\delta(t)\|^2$ is converged to the set $\mathcal{C}_2$. Comparing Figs. 7.19 and 7.14, it can be seen that the impact of replacement attacks on the consensus of MASs in this example is greater than that of injection attacks in Example 7.3. So it is necessary to apply more impulsive control to ensure the consensus of MASs in the sense of mean square. Therefore, in the simulation of Example 7.4, the number of impulses is set to 200 (only 30 in Example 7.3).

In this example, the relationship between the convergence rate r and the upper bound ρ_2 of the impulsive interval h_2 and the error bound are shown in Figs. 7.21 and 7.22, respectively. As can be seen from Fig. 7.22, if the convergence rate r is too small, the obtained error bounded ρ_2 is large. Therefore, the convergence rate r is still an important control parameter in the case of MASs under replacement attacks. It is necessary to choose the appropriate convergence rate according to whether one wants to obtain a larger impulsive interval or a smaller error bound.

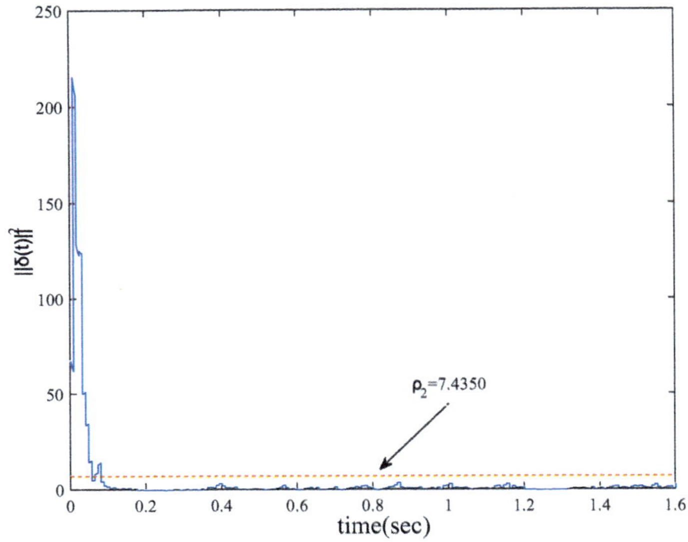

Fig. 7.20 The curves of $\|\delta(t)\|^2$ and error bound in Example 7.4

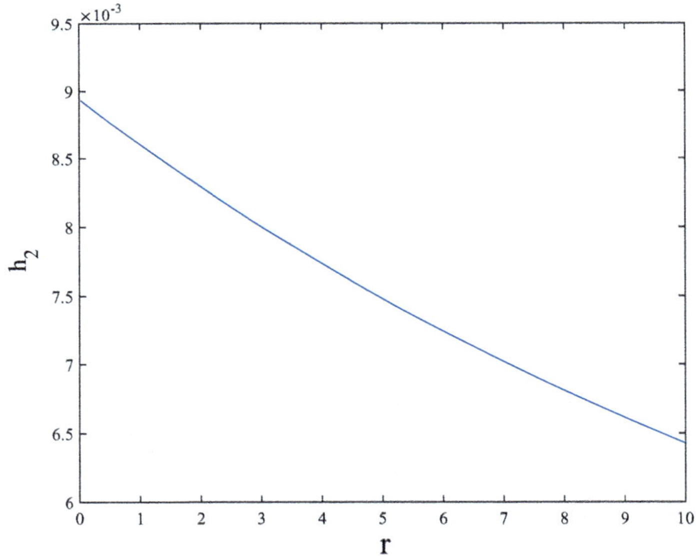

Fig. 7.21 The relationship of convergence rate r and h_2 in Example 7.4

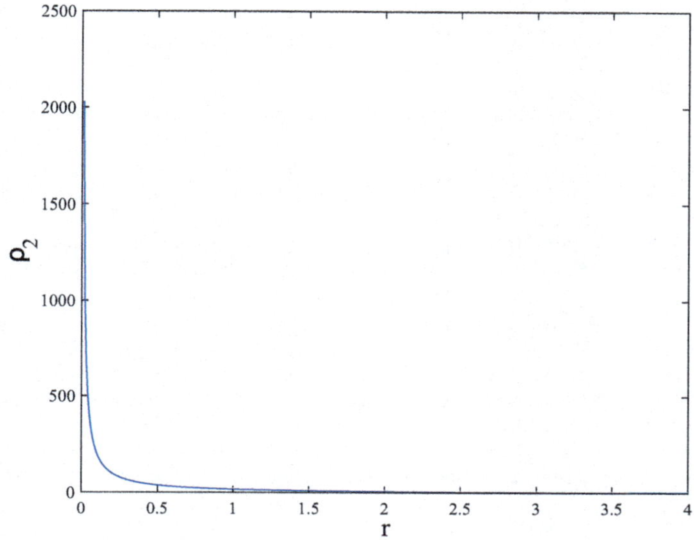

Fig. 7.22 The relationship of r and ρ_2 in Example 7.4

7.5 Summary

This chapter investigates the impulsive consensus control problem of MASs under cyber attacks. (1) This chapter first focuses on the situation where the system is subjected to DoS attacks. Considering that DoS attacks can interrupt the information transmission of some edges in the system, three different windows are introduced, and appropriate distributed impulsive controllers are designed. The control strategies in each of the three windows are provided to ensure the consensus of MASs under DoS attacks. (2) Next, this chapter considers two cases in which the system is subjected to injection attacks and replacement attacks in the controller-to-actuator channel. For these two types of deception attacks, sufficient conditions for achieving consensus in mean-squared sense are obtained, and the relationships between impulsive control gain, lower and upper limits of impulsive interval, convergence speed, and selection rules for these control parameters are given to achieve the best control effect. The numerical simulation results have demonstrated the correctness of the theoretical results in this chapter and present its superiority over control strategies in other existing results.

References

Bu X, Yu W, Yin Y, Zhou Z (2023) Event-triggered data-driven control for nonlinear systems under frequency-duration-constrained DoS attacks. IEEE Trans Inf Forensics Secur 18:1449–1460

Chen WH, Wan Q, Lu X (2023) Guaranteed performance impulsive tracking control of multi-agents systems under discrete-time deception attacks. Commun Nonlinear Sci Numer Simul 117:106905

Ding L, Li J, Ye M, Zhao Y (2022) Fully distributed resilient cooperative control of vehicular platoon systems under DoS attacks. IEEE/CAA J Autom Sinica 9(5):937–940

Guo X, Li Q, Ji L, Wang J (2023) Quasi-synchronization of heterogeneous nonlinear multi-agent systems subject to DOS attacks with impulsive effects. Syst Control Lett 173:105463

He W, Gao X, Zhong W, Qian F (2018) Secure impulsive synchronization control of multi-agent systems under deception attacks. Inf Sci 459:354–368

He W, Mo Z, Han QL, Qian F (2020) Secure impulsive synchronization in Lipschitz-type multi-agent systems subject to deception attacks. IEEE/CAA J Autom Sinica 7(5):1326–1334

Li F, Zhou Z (2023) Distributed model-free adaptive control for MIMO nonlinear multiagent systems under deception attacks. IEEE Trans Syst Man Cybern: Syst 53(4):2281–2291

Li J, Ren CE, Chen CLP (2022) Adaptive impulsive security consensus control of uncertain multi-agent systems with randomly occurring nonlinearities. Int J Robust Nonlinear Control 32(18):9849–9868

Liu Z, Shi Y, Yan H, Han B, Guan Z (2023) Secure consensus of multi-agent systems via impulsive control subject to deception attacks. IEEE Trans Circuits Syst II Express Briefs 70(1):166–170

Wang Y, He W (2019) Impulsive consensus of leader-following nonlinear multi-agent systems under DoS attacks. Ind Electr Conf 6274–6279

Wen G, Zhai X, Peng Z, Rahmani A (2020) Fault-Tolerant secure consensus tracking of delayed nonlinear multi-agent systems with deception attacks and uncertain parameters via impulsive control. Commun Nonlinear Sci Numer Simul 82:105043

Ye D, Shao Y (2019) Quasi-synchronization of heterogeneous nonlinear multi-agent systems subject to DOS attacks with impulsive effects. Neurocomputing 366:131–139

Zhang X, Zhang H, Yang X, Yan H (2023) Distributed dimensionality reduction diltering for CPSs under DoS attacks. IEEE/CAA J Autom Sinica 10(4):1080–1082

Chapter 8
Distributed Delayed Impulsive Formation of Second-Order Multi-agent Systems with Switching Topologies

8.1 Introduction

In the field of cooperative control of multi-agent systems (MASs), formation control plays an important role. Its research goal is to drive multiple agents in MASs to achieve an ideal formation. The formation problem of MASs with a leader plays an important role in real life. In the study of formation control on MASs, because the impulsive control only transmits information at a certain instant, the communication redundancy is significantly reduced, and the confidentiality of information transmission is greatly improved. Therefore, the introduction of impulsive protocol becomes a very efficient and practical method to realize the formation of MASs.

Moreover, the sudden transmission of interruption also exists; as a result, the communication topology of MASs may be changed due to some undesired faults. Switching topology means that the topology will change with certain regular switching logic. Sun et al. (2020) adopted a sampling interval discretization method for discrete-time switched polytopic system. Chen et al. (2021b) addressed a consensus problem for discrete-time dynamical networks with switching topology and time delays under actuator saturation for the first time. An et al. (2023) realized the tracking of MASs with switching topologies via a distributed cycle stabilization approach. Guo et al. (2023a) achieved the mean-squared output consensus of MASs under randomly switching topologies and time delays. Fan et al. (2024) investigated the optimal topology switching problem for consensus of MASs. Guo et al. (2023b) realized the output containment of MASs subjected to Markovian switching topologies and distributed delays.

Recently, there are some researches about impulsive coordination of systems under switching topology. Wang et al. (2024) concerned with the exponential consensus issue for MASs of impulsive partial differential equations with switching topology. Yaghoubi and Talebi (2020) considered a cluster consensus problem for fractional-order nonlinear MASs with fixed and switching topologies and

Z. Zhang et al., *Distributed Impulsive Coordination of Multi-Agent Systems*, Intelligent Control and Learning Systems 15, https://doi.org/10.1007/978-981-97-6262-0_8

time delays. Hu et al. (2022) investigated the hybrid event-triggered and impulsive consensus problems for leaderless and leader-following MASs with switching topologies. Yu et al. (2023) addressed the consensus problem of impulsive control for the MASs with partially unknown transition probabilities under uncertain semi-Markovian switching topologies. Wang et al. (2023) focused on a finite-time consensus of nonlinear MASs by state-constraint impulsive control protocol. Chen et al. (2021a) proposed an event-triggering impulsive control strategy for coordination of systems with switching topology, in which the impulsive time sequence can be different from the event-triggered time sequence. Ma et al. (2021) investigated the impulsive consensus problem of the one-sided Lipschitz nonlinear MASs with Semi-Markov switching topologies. Yang et al. (2020) investigated global exponential synchronization almost surely of complex networks with node delay and switching topology. It's more practical to consider switching topology in the formation control of MASs. In this chapter, both switching topology and time delays are considered for a class of second-order MASs.

Time delay is often a source of instability that occurs in many engineering systems. It is necessary to consider control systems with input delay. Control inputs are often subject to delays in sensor-to-controller and controller-to-actuator channels. If the control delay is not considered in the design of control protocol, the existence of the control delay may lead to the serious deterioration or even instability of the controlled system performance (Huang and Xu 2023). In recent years, there are some works about impulsive coordination of MASs with input delay. Zhou et al. (2014a) investigated the containment control of MASs with input delay using impulsive algorithms without using velocity measurements and researched the consensus tracking problem for second-order MASs with input delays in Zhou et al. (2014b). Liu et al. (2022) focused on the problem of formation tracking for MASs via impulsive control subject to input time delay.

Based on the preceding discussions, this chapter realizes the formation of second-order MASs with switching topologies and time delay via the impulsive protocol. It should be noted that both time delays and switching topologies are considered simultaneously in this chapter, which is more flexible and practical in practical application.

8.2 Problem Description

Suppose that the MASs with N identical agents is described by

$$\begin{cases} \dot{x}_i(t) = v_i(t), \\ \dot{v}_i(t) = u_i(t), \end{cases} \tag{8.1}$$

where $x_i = [x_{i1}, x_{i2}, \ldots, x_{in}]^T \in R^n$ is the position of node i and $v_i = [v_{i1}, v_{i2}, \ldots, v_{in}]^T \in R^n$ is the velocity of node i ($i = 1, 2, \ldots, N$).

Consider the dynamics of the leaders as

$$\begin{cases} \dot{x}_0(t) = v_0(t), \\ \dot{v}_0(t) = 0, \end{cases} \tag{8.2}$$

where $x_0 = [x_{01}, x_{02}, \ldots, x_{0n}{}^T] \in R^n$ is the position of leader and $v_0 = [v_{01}, v_{02}, \ldots, v_{0n}]^T \in R^n$ is the constant velocity of the leader. It is assumed that $\tau(t)$ is the time delay which satisfies $\tau(t) = rT + \tau_0(t)$ and $0 \le \tau_0(t) < T$, where $T = t_k - t_{k-1}$ is sampling period and r is a nonnegative integer.

8.3 Main Results

8.3.1 Delay Impulsive Controller Design

The impulsive control input of agent i is designed as

$$\begin{aligned} &u_i(t) \\ &= \Bigg\{ \sum_{k=1}^{\infty} b(\sum_{j\in N_i} a_{ij}(x_i(t-\tau(t)) - h_i - x_j(t-\tau(t)) \\ &+h_j) + c_i(x_i(t-\tau(t)) - h_i - x_0(t-\tau(t)))) \\ &+l(\sum_{j\in N_i} a_{ij}(v_i(t-\tau(t)) - v_j(t-\tau(t))) + c_i(v_i(t-\tau(t)) - v_0)) \Bigg\} \delta^*(t-t_k), \end{aligned} \tag{8.3}$$

where $k \in N^+, i = 1, 2, \ldots, N$ and $b, l < 0 \in R$ is the impulsive control gain of the system, and $\delta^*(t)$ is the Dirac delta function which satisfies $\delta^*(t) = 0$ for $t \neq 0$ and $\int_{-\infty}^{+\infty} \delta^*(t) = 1$. $\delta^*(t)$ has the fundamental property that $\int_{a-\varepsilon}^{a+\varepsilon} \delta^*(t-a) f(t) = f(a)$ when $\varepsilon \neq 0$, h_i is a time-invariant vector which is a desired displacement between every follower and leader.

The objective of the formation problem can be described as

$$\lim_{t\to+\infty} \|x_i(t) - x_0(t) - h_i\| = 0, \tag{8.4}$$

$$\lim_{t\to+\infty} \|v_i(t) - v_0\| = 0, \tag{8.5}$$

for $0 \le \tau_0(t) < T$, $t_{k-1} \le t_k - \tau_0(t) < t_k$ and $x_i(t_k - \tau_0(t)) = x_i(t_{k-1}) + (T - \tau_0(t))\, v_i\left(t_{k-1}^+\right) = x_i(t_{k-1}) + (T - \tau_0(t))\, v_i(t_k)$, $\lim\limits_{t\to t_k^-} v_i(t) = v_i(t_k)$ means that $v_i(t)$ is left continuous at time t_k.

Applying the impulsive control (8.3) to system (8.1), it yields

$$\begin{cases} \dot{x}_i(t) = v_i(t), \\ \dot{v}_i(t) = 0, \quad t \ne t_k, \\ \Delta v_i(t_k) = v_i\left(t_k^+\right) - v_i\left(t_k^-\right) \\ \qquad = b(\sum\limits_{j\in N_i} a_{ij}(x_i(t_{k-1}) - h_i - x_j(t_{k-1}) + h_j) \\ \qquad +c_i(x_i(t_{k-1}) - h_i - x_0(t_{k-1}))) \\ \qquad +(l + b(T - \tau_0(t))) \\ \qquad\quad (\sum\limits_{j\in N_i} a_{ij}(v_i(t_k) - v_j(t_k)) + c_i(v_i(t_k) - v_0)), \\ v(t_0^+) = v(t_0), \end{cases} \tag{8.6}$$

where $\Delta v_i(t_k)$ is the jump of the velocity of follower agent i at switching instant t_k, and b, l are the impulsive control gain.

Remark 8.1 It is easy to observe that formula (8.6) ignores the time delay of velocity. Assume that $0 \le \tau_0(t) < T$ and the velocity keeps a constant when $t \ne t_k$. Therefore, it doesn't matter if there exists a time delay for velocity.

Denote $\delta_i(t) = x_i(t) - h_i - x_0(t)$, $\mu_i(t) = v_i(t) - v_0$.

Definition 8.1 The error for node i is defined as

$$e_i(t) = \begin{bmatrix} \delta_i(t) \\ \mu_i(t) \end{bmatrix}, i = 1, \ldots, N. \tag{8.7}$$

When $t = t_k^+$

$$
\begin{aligned}
e_i\left(t_k^+\right) &= \begin{bmatrix} \delta_i\left(t_k^+\right) \\ \mu_i\left(t_k^+\right) \end{bmatrix} \\
&= \begin{bmatrix} \delta_i\left(t_k^-\right) \\ v_i\left(t_k^+\right) - v_0 \end{bmatrix} \\
&= \begin{bmatrix} \delta_i\left(t_k^-\right) \\ \Delta v_i\left(t_k\right) + v_i\left(t_k^-\right) - v_0 \end{bmatrix} \\
&= \begin{bmatrix} \delta_i\left(t_k^-\right) \\ b \sum\limits_{j \in N_i} \tilde{l}_{ij} \delta_i\left(t_k\right) + \kappa \sum\limits_{j \in N_i} \tilde{l}_{ij} \mu_i\left(t_k\right) + \mu_i\left(t_k\right) \end{bmatrix},
\end{aligned}
\tag{8.8}
$$

where $\tilde{l}_{ij}$ is the element of $\tilde{L}$, $\tilde{L} = L_\sigma + C_\sigma$, $\kappa = l + b\left(T - \tau_0\left(t\right)\right)$.

Formula (8.8) can be compacted as

$$
e(t_k^+) = \begin{bmatrix} \delta(t_k) \\ b\boldsymbol{L}\delta(t_{k-1}) + \kappa\boldsymbol{L}\mu(t_k) + \mu(t_k) \end{bmatrix},
\tag{8.9}
$$

where $\boldsymbol{L} = \tilde{L} \otimes I_n$, $\mu\left(t_k\right) = \left(v\left(t_k\right) - v_0 \otimes 1_N\right)$, $\delta\left(t_k\right) = x\left(t_k\right) - x_0\left(t_k\right) \otimes 1_N - H$, $H = \{h_1, h_2, \ldots, h_i\}^T$, $v(t_k) = \{v_1(t_k), v_2(t_k), \ldots, v_i(t_k)\}^T$, $x(t_k) = \{x_1(t_k), x_2(t_k), \ldots, x_i(t_k)\}^T$, $e(t_k^+) = \left\{e_1(t_k^+), e_2(t_k^+), \ldots, e_i(t_k^+)\right\}^T$.

When $t = t_{k+1}$

$$
\begin{aligned}
e\left(t_{k+1}\right) &= \begin{bmatrix} \delta\left(t_{k+1}\right) \\ \mu\left(t_{k+1}\right) \end{bmatrix} \\
&= \begin{bmatrix} \delta\left(t_k^+\right) + T\mu\left(t_k^+\right) \\ \mu\left(t_k^+\right) \end{bmatrix}.
\end{aligned}
\tag{8.10}
$$

From (8.9) and (8.10), one gets

$$
e\left(t_{k+1}\right) = \left(\begin{bmatrix} 1 & T \\ 0 & 1 \end{bmatrix} \otimes I_{nN}\right) e\left(t_k^+\right).
\tag{8.11}
$$

Note that

$$
\mu\left(t_k^+\right) = b\boldsymbol{L}\delta\left(t_{k-1}\right) + \kappa\boldsymbol{L}\mu\left(t_k\right) + \mu\left(t_k\right).
\tag{8.12}
$$

From (8.11), one has

$$
\begin{aligned}
e\left(t_{k+1}\right) &= \left(\begin{bmatrix} 1 & T \\ 0 & 1 \end{bmatrix} \otimes I_{nN}\right) e(t_k) + \begin{bmatrix} T\mu(t_k^+) - T\mu(t_k) \\ \mu(t_k^+) - \mu(t_k) \end{bmatrix} \\
&= \left(\begin{bmatrix} 1 & T \\ 0 & 1 \end{bmatrix} \otimes I_{nN}\right) e(t_k) + \begin{bmatrix} Tb\boldsymbol{L} & T\kappa\boldsymbol{L} \\ b\boldsymbol{L} & \kappa\boldsymbol{L} \end{bmatrix} \begin{bmatrix} \delta\left(t_{k-1}\right) \\ \mu\left(t_k\right) \end{bmatrix} \\
&= \left(\begin{bmatrix} 1 & T \\ 0 & 1 \end{bmatrix} \otimes I_{nN} + \begin{bmatrix} 0 & T\kappa\boldsymbol{L} \\ 0 & \kappa\boldsymbol{L} \end{bmatrix}\right) e\left(t_k\right) + \begin{bmatrix} Tb\boldsymbol{L} & 0 \\ b\boldsymbol{L} & 0 \end{bmatrix} e\left(t_{k-1}\right)
\end{aligned}
\tag{8.13}
$$

and then

$$e(t_{k+1}) = (M_1 \otimes I_n)\, e(t_k) + (M_2 \otimes I_n)\, e(t_{k-1}), \tag{8.14}$$

where

$$M_1 = \begin{bmatrix} 1 & T \\ 0 & 1 \end{bmatrix} \otimes I_N + \begin{bmatrix} 0 & T\kappa\tilde{L} \\ 0 & \kappa\tilde{L} \end{bmatrix}, \tag{8.15}$$

$$M_2 = \begin{bmatrix} Tb\tilde{L} & 0 \\ b\tilde{L} & 0 \end{bmatrix}. \tag{8.16}$$

Let $\eta(k) = \left[\, e(t_k) \;\; e(t_{k-1}) \,\right]^T$, one gets an augmented system

$$\eta(k+1) = Q\eta(k), \tag{8.17}$$

where

$$Q = \begin{bmatrix} M_1 & M_2 \\ I_{2N} & 0 \end{bmatrix}. \tag{8.18}$$

In this subsection, under the time-delayed condition, a portable impulsive control law is designed and the switching topologies are considered as well. In the next subsection, some necessary lemma and theorem are proposed.

8.3.2 Convergence Analysis

In this subsection, the main results will be given to realize the formation of second-order MASs.

Lemma 8.1 (Liu et al. 2013) *If the spectral radius of Q has $\rho(Q) < 1$, the formation of MASs (8.6) can be realized.*

Remark 8.1 Lemma 8.1 gives a necessary and sufficient condition, but it does not show what kind of communication graph can reach formation, and how to determine suitable control gains and sampling period T. The following theorem addresses these issues.

Theorem 8.1 *Suppose that Assumption 1.1 holds, the impulsive interval T and time delay τ satisfy*

$$T < \min \frac{-4\mathrm{Re}(\lambda_i)|\lambda_i|^2 - 2\zeta|\lambda_i|^4}{\frac{4b}{\zeta^2}\mathrm{Im}^2(\lambda_i) + b|\lambda_i|^4}, \tag{8.19}$$

where λ_i are the eigenvalues of $\tilde{L}$, $i = 1, 2, 3, \ldots, N$, $\zeta = l - b\tau_0(t)$, the MASs (8.6) with algorithm (8.3) can achieve the formation.

Proof Let λ be the eigenvalue of Q. Then one has

$$\begin{aligned}
&\det(\lambda I_{4N} - Q) \\
&= \det\left(\lambda I_{4N} - \begin{bmatrix} M_1 & M_2 \\ I_{2N} & 0 \end{bmatrix}\right) \\
&= \det(\lambda^2 I_{2N} - \lambda M_1 - M_2) \\
&= \begin{bmatrix} \lambda^2 I_N - \lambda I_N - Tb\tilde{L} & -\lambda T I_N - \lambda T\kappa\tilde{L} \\ -b\tilde{L} & \lambda^2 I_N - \lambda I_N - \lambda\kappa\tilde{L} \end{bmatrix} \\
&= \prod_{i=1}^{N} \left(\lambda^4 - (2+\kappa\lambda_i)\,\lambda^3 + (1+\zeta\lambda_i)\,\lambda^2\right).
\end{aligned} \tag{8.20}$$

Let

$$f_i(\lambda) = \lambda^2 - (2+\kappa\lambda_i)\,\lambda + 1 + \zeta\lambda_i, \tag{8.21}$$

$i = 1, 2, 3, \ldots, N$. $\rho(Q) < 1$ if and only if $f_i(\lambda)$ is Schur stable, then the consensus of system is achieved. Applying a bilinear transformation $s = \frac{\lambda+1}{\lambda-1}$ on $f_i(\lambda)$, then

$$\begin{aligned}
\varphi_i(s) &= (s-1)^2 f_i\left(\frac{s+1}{s-1}\right) \\
&= (-b\lambda_i T)\,s^2 - 2\zeta\lambda_i s + 4 + 2\lambda_i\zeta + b\lambda_i T.
\end{aligned} \tag{8.22}$$

Let

$$\varphi_i^{'}(s) = s^2 + \frac{2\zeta}{bT}s + \frac{4 + 2\lambda_i\zeta + b\lambda_i T}{-b\lambda_i T}. \tag{8.23}$$

Complex polynomial $f_i(\lambda)$ is Schur stable if and only if the complex polynomial $\varphi_i^{'}(s)$ is Hurwitz stable.

Let

$$\varphi_i^{'}(s) = m(\omega) + jn(\omega), \tag{8.24}$$

where $m(\omega)$ and $n(\omega)$ are the real value and imaginary value of $\varphi_i^{'}(s)$, respectively.

Let $s = j\omega$, then

$$\begin{aligned}
\varphi_i^{'}(j\omega) &= -\omega^2 + \frac{2\zeta}{bT}j\omega + \frac{4 + 2\lambda_i\zeta + b\lambda_i T}{-b\lambda_i T} \\
&= m(\omega) + jn(\omega),
\end{aligned} \tag{8.25}$$

where

$$m(\omega) = -\omega^2 - \frac{4}{bT}\mathrm{Re}\left(\frac{1}{\lambda_i}\right) - \frac{2\zeta + bT}{bT}, \tag{8.26}$$

$$p(\omega) = \frac{2\zeta}{bT}\omega - \frac{4}{bT}\operatorname{Im}\left(\frac{1}{\lambda_i}\right). \tag{8.27}$$

The complex polynomial $\varphi_i^{'}(s)$ is Hurwitz stable if the following conditions hold:
(a). $m(\omega) = 0$ has two different real roots $m_1 < m_2$;
(b). $m_1 < p < m_2$, where n_1 is the unique root of $p(\omega) = 0$;
(c). $m(0)\,\dot{p}(0) - \dot{m}(0)\,p(0) > 0$.
From the above conditions and (8.24)–(8.25), it yields

$$\begin{cases} \Delta_m = -\frac{16}{bT}\operatorname{Re}\left(\frac{1}{\lambda_i}\right) - \frac{8\zeta + 4bT}{bT}, \\ m_1 = -\frac{\sqrt{\Delta_m}}{2} = -\sqrt{-\frac{4}{bT}\operatorname{Re}\left(\frac{1}{\lambda_i}\right) - \frac{2\zeta + bT}{bT}}, \\ m_2 = \frac{\sqrt{\Delta_m}}{2} = \sqrt{-\frac{4}{bT}\operatorname{Re}\left(\frac{1}{\lambda_i}\right) - \frac{2\zeta + bT}{bT}}, \\ p = \frac{2}{\zeta}\operatorname{Im}\left(\frac{1}{\lambda_i}\right). \end{cases} \tag{8.28}$$

According to conditions (a) and (b), the results can be expressed as

$$\begin{cases} -\frac{16}{bT}\operatorname{Re}\left(\frac{1}{\lambda_i}\right) - \frac{8\zeta+4bT}{bT} > 0, \\ -\sqrt{-\frac{4}{bT}\operatorname{Re}\left(\frac{1}{\lambda_i}\right) - \frac{2\zeta+bT}{bT}} < \frac{2}{\zeta}\operatorname{Im}\left(\frac{1}{\lambda_i}\right) \\ < \sqrt{-\frac{4}{bT}\operatorname{Re}\left(\frac{1}{\lambda_i}\right) - \frac{2\zeta+bT}{bT}}. \end{cases} \tag{8.29}$$

It follows that

$$\begin{cases} T < -\frac{2\zeta|\lambda_i|^2+4\operatorname{Re}(\lambda_i)}{b|\lambda_i|^2}, \\ T < \frac{-4\operatorname{Re}(\lambda_i)|\lambda_i|^2-2\zeta|\lambda_i|^4}{\frac{4b}{\zeta^2}\operatorname{Im}^2(\lambda_i)+b|\lambda_i|^4}. \end{cases} \tag{8.30}$$

Thus

$$T < \min\left\{\min\left(\frac{-4\operatorname{Re}(\lambda_i)|\lambda_i|^2 - 2\zeta|\lambda_i|^4}{\frac{4b}{\zeta^2}\operatorname{Im}^2(\lambda_i) + b|\lambda_i|^4}\right),\right. \left.\min\left(\frac{-2\zeta|\lambda_i|^2 - 4\operatorname{Re}(\lambda_i)}{b|\lambda_i|^2}\right)\right\}. \tag{8.31}$$

According to condition (c), it can be expressed as

$$\left(-\frac{4}{Tb}\operatorname{Re}\left(\frac{1}{\lambda_i}\right) - 1 - \frac{2\zeta}{Tb}\right)\left(\frac{2\zeta}{Tb}\right) > 0. \tag{8.32}$$

It follows that

$$T < \min\left(\frac{-4\mathrm{Re}(\lambda_i)|\lambda_i|^2 - 2\zeta|\lambda_i|^4}{b|\lambda_i|^4}\right). \tag{8.33}$$

From (8.31) and (8.33), it concludes that

$$T < \min \frac{-4\mathrm{Re}(\lambda_i)|\lambda_i|^2 - 2\zeta|\lambda_i|^4}{\frac{4b}{\zeta^2}\mathrm{Im}^2(\lambda_i) + b|\lambda_i|^4}. \tag{8.34}$$

Thus, the proof of Theorem 8.1 is hence completed.

Remark 8.2 From (8.19) in Theorem 8.1, one gets the range of impulsive interval T. It should be noted that a large impulsive interval can reduce the conservative of the control method. In order to reduce the conservatism of impulsive protocol while ensuring the realization of formation, the impulsive interval T can be selected as

$$T = \min \frac{-4\mathrm{Re}(\lambda_i)|\lambda_i|^2 - 2\zeta|\lambda_i|^4}{\frac{4b}{\zeta^2}\mathrm{Im}^2(\lambda_i) + b|\lambda_i|^4} - \epsilon,$$

where ϵ is a small constant.

Corollary 8.1 *If there exists a desire condition that the communication delay between any neighbors can be ignored which means that the time delay* $\tau = 0$*. From Theorem 8.1, it concludes that if there exists* T *which satisfies*

$$T < \min \frac{-4\mathrm{Re}(\lambda_i)|\lambda_i|^2 - 2l|\lambda_i|^4}{\frac{4b}{l^2}\mathrm{Im}^2(\lambda_i) + b|\lambda_i|^4}, \tag{8.35}$$

where λ_i *are the eigenvalues of* $\tilde{L}$*, the MASs (8.6) with algorithm (8.3) can achieve formation.*

It should be noted that the proof of Corollary 8.1 is almost same as Theorem 8.1, so it is omitted here.

Remark 8.3 From Corollary 8.1, it can be seen that the absence of input delay is only a special case of Theorem 8.1. Due to the factors such as actual environment and transmission distance, the given control input often has delays, so the distributed impulsive control input also needs to consider the impact of time delay. Therefore, considering the existence of input delay of impulsive control is more satisfied with the actual situation.

8.4 Simulation Results

In this section, two simulation examples are provided to show the effectiveness of the theoretical results obtained in this chapter. Consider the MASs consisting of one leader and four followers.

Example 8.1 Consider the MASs with fixed topology $\mathcal{G}_1$ and the corresponding system matrices are given as follows:

$$A_1 = \begin{bmatrix} 0\,1\,0\,0 \\ 1\,0\,1\,0 \\ 0\,1\,0\,1 \\ 0\,0\,1\,0 \end{bmatrix},\ D_1 = \begin{bmatrix} 1\,0\,0\,0 \\ 0\,2\,0\,0 \\ 0\,0\,2\,0 \\ 0\,0\,0\,1 \end{bmatrix},$$

$$C_1 = \begin{bmatrix} 1\,0\,0\,0 \\ 0\,0\,0\,0 \\ 0\,0\,0\,0 \\ 0\,0\,0\,0 \end{bmatrix},\ L_1 = D_1 - A_1 = \begin{bmatrix} 1 & -1 & 0 & 0 \\ -1 & 2 & -1 & 0 \\ 0 & -1 & 2 & -1 \\ 0 & 0 & -1 & 1 \end{bmatrix},$$

then one gets $L_1 + C_1 = \begin{bmatrix} 2 & -1 & 0 & 0 \\ -1 & 2 & -1 & 0 \\ 0 & -1 & 2 & -1 \\ 0 & 0 & -1 & 1 \end{bmatrix}$.

The topology of the communication graph $\mathcal{G}_1$ is presented in Fig. 8.1. It's easy to calculate the eigenvalues of $L_1 + C_1$ as $\lambda_1 = 0.1206$, $\lambda_2 = 1.0000$, $\lambda_3 = 2.3473$, $\lambda_4 = 3.5321$. Let the time delay $\tau_0(t) = 0.01$. Thus, according to Theorem 8.1, the system can achieve formation behavior if

$$T < \min \frac{-4\mathrm{Re}(\lambda_i)|\lambda_i|^2 - 2\zeta|\lambda_i|^4}{\frac{4b}{\zeta^2}\mathrm{Im}^2(\lambda_i) + b|\lambda_i|^4} = 0.1525.$$

Let $T = 0.1425$, impulsive gain as $b = -1$, $l = -0.5$. Design the desired displacement as $h_1 = [15, 10]^T$, $h_2 = [15, -10]^T$, $h_3 = [-15, -10]^T$, $h_4 = [-15, 10]^T$. The position and velocity errors of all agents are shown in Figs. 8.2 and 8.3, and the trajectories are shown in Fig. 8.4.

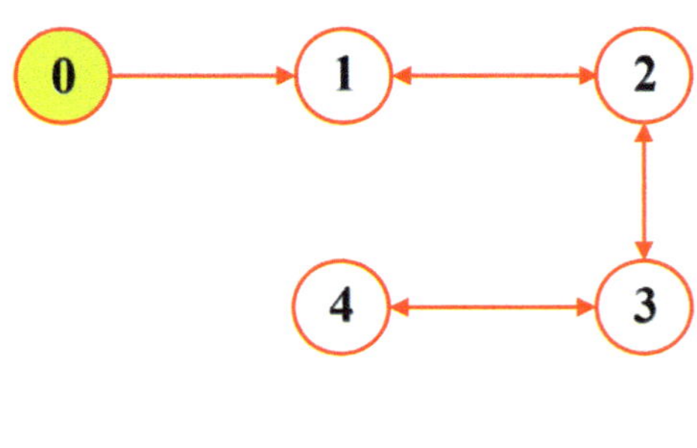

Fig. 8.1 Topology of the communication graph $\mathcal{G}_1$

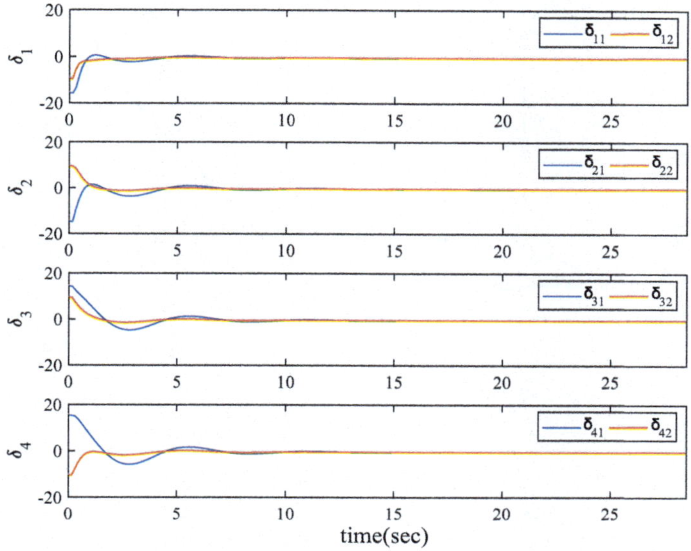

Fig. 8.2 Position errors δ_{xi} of leader–follower system with topology of graph $\mathcal{G}_1$

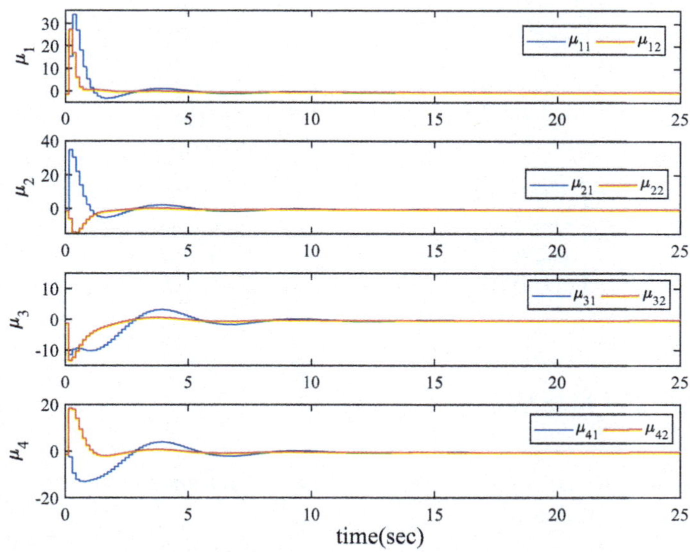

Fig. 8.3 Velocity errors μ_i of leader–follower system with topology of graph $\mathcal{G}_1$

Example 8.2 Consider the MASs with switching topologies. Suppose that the topologies are expressed as $\mathcal{G}_1$, $\mathcal{G}_2$, where $\mathcal{G}_1$ is the same with that in Example 8.1 and the corresponding system matrices of graph $\mathcal{G}_2$ are given as follows:

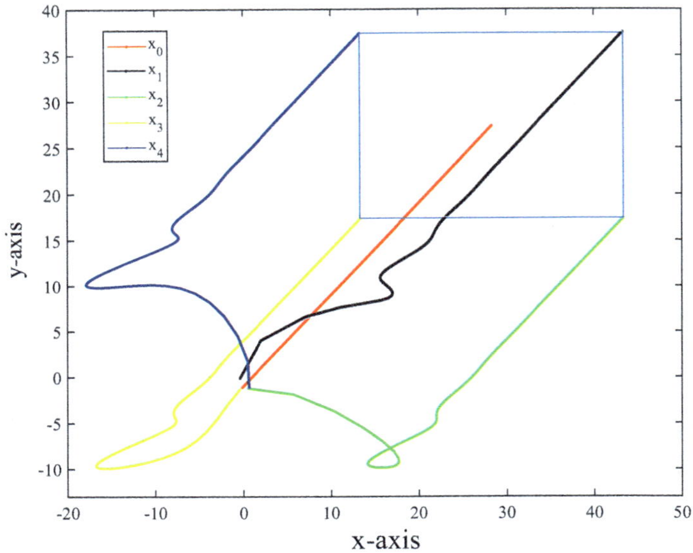

Fig. 8.4 The trajectories of the leader and four followers with topology of graph $\mathcal{G}_1$

$$A_2 = \begin{bmatrix} 0\,0\,0\,1 \\ 1\,0\,0\,0 \\ 0\,1\,0\,0 \\ 0\,0\,1\,0 \end{bmatrix}, \ D_2 = \begin{bmatrix} 1\,0\,0\,0 \\ 0\,1\,0\,0 \\ 0\,0\,1\,0 \\ 0\,0\,0\,1 \end{bmatrix},$$

$$C_2 = \begin{bmatrix} 1\,0\,0\,0 \\ 0\,0\,0\,0 \\ 0\,0\,0\,0 \\ 0\,0\,0\,0 \end{bmatrix}, L_2 = D_2 - A_2 = \begin{bmatrix} 1 & 0 & 0 & -1 \\ -1 & 1 & 0 & 0 \\ 0 & -1 & 1 & 0 \\ 0 & 0 & -1 & 1 \end{bmatrix},$$

and then one gets $L_2 + C_2 = \begin{bmatrix} 2 & 0 & 0 & -1 \\ -1 & 1 & 0 & 0 \\ 0 & -1 & 1 & 0 \\ 0 & 0 & -1 & 1 \end{bmatrix}$.

The topology of the communication graph $\mathcal{G}_2$ is presented in Fig. 8.5. It is easy to calculate the eigenvalues of $L_2 + C_2$ as $\lambda_1 = 2.3803, \lambda_2 = 1.2194 + 0.9145j, \lambda_3 = 1.2194 - 0.9145j, \ \lambda_4 = 0.1808$. One defines a switching signal $\sigma : [0, +\infty) \to \{\mathcal{G}_1, \mathcal{G}_2\}$. In time intervals $[t_{4k}, t_{4k+1}), \ k \in \mathbb{N}^+$. The adjacency matrices are A_1 and C_1. In other time intervals, the adjacency matrices are A_2 and C_2. The time delay τ, control gains b and l and the displacement h_i are chosen as same as Example 8.1.

According to Theorem 8.1, the system can achieve formation behavior if

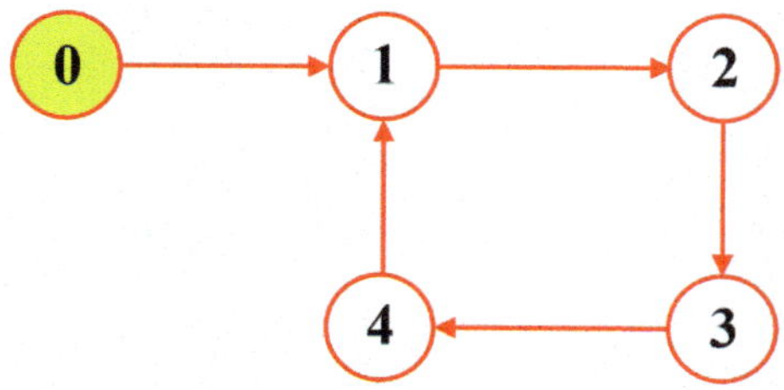

Fig. 8.5 Topology of the communication graph $\mathcal{G}_2$

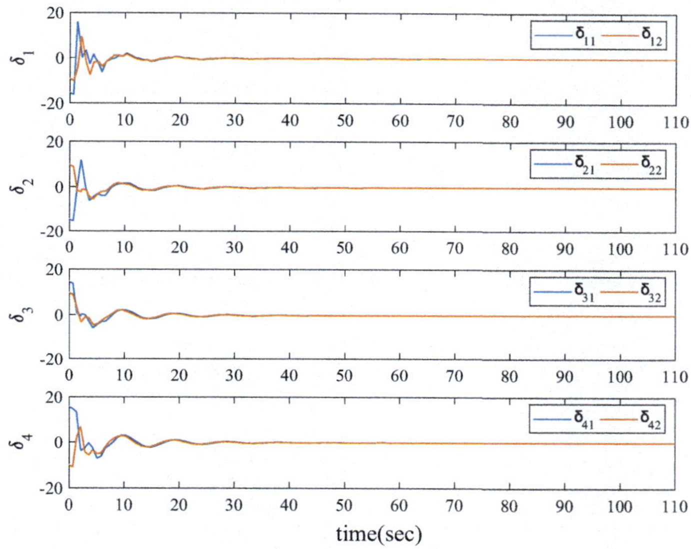

Fig. 8.6 Position errors δ_{xi} of leader–follower system with switching topologies

$$T < \min \frac{-4\text{Re}(\lambda_i)|\lambda_i|^2 - 2\zeta|\lambda_i|^4}{\frac{4b}{\zeta^2}\text{Im}^2(\lambda_i) + b|\lambda_i|^4} = 0.3126.$$

Let $T = 0.30265$ uhder graph $\mathcal{G}_2$, $T = 0.1425$ under graph $\mathcal{G}_1$. The position and velocity errors of all agents are shown in Figs. 8.6 and 8.7, and the trajectories are shown in Fig. 8.8.

Throughout the chapter, the initial conditions of position and velocity for every agent are given as $x_0 = [-0.2, -1]^T$, $x_1 = [-0.4, -0.1]^T$, $x_2 = [0.5, -1.1]^T$, $x_3 = [-0.5, -1.2]^T$, $x_4 = [0.6, -1]^T$, $v_0 = [1, 1]^T$, and $v_1 = v_2 = v_3 = v_4 = [0, 0]^T$, respectively.

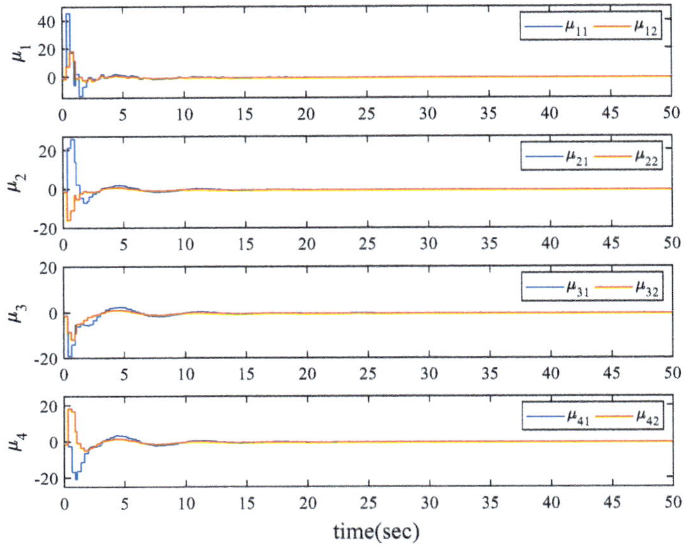

Fig. 8.7 Velocity errors μ_i of leader–follower system with switching topologies

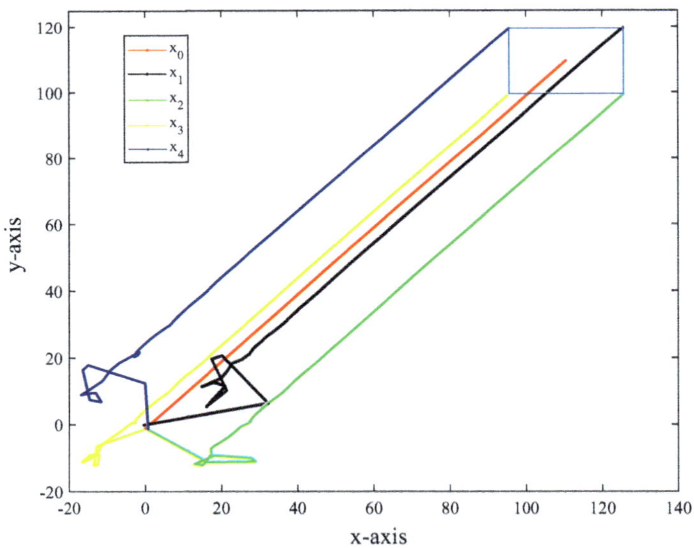

Fig. 8.8 Trajectories of the leader and four followers with switching topologies

8.5 Summary

In this chapter, a time-delayed impulsive protocol is proposed to achieve the formation of MASs with switching topologies. Based on the theories of impulsive differential equation and discrete-time systems, a sufficient condition is obtained. From the simulation part, it draws a conclusion that four followers will complete a geometric formation of a rectangle with this control law eventually.

References

An S, Lian J, Wang D (2023) Distributed tracking control of multi-agent systems with transfer restricted switching topologies: a cycle stabilization approach. IEEE Trans Netw Sci Eng 10(4):1964–1974

Chen J, Chen B, Zeng Z (2021a) Synchronization in multiple neural networks with delay and disconnected switching topology via event-triggered impulsive control strategy. IEEE Trans Industr Electron 68(3):2491–2500

Chen Y, Wang Z, Hu J, Han QL (2021b) Synchronization control for discrete-time-delayed dynamical networks with switching topology under actuator saturations. IEEE Trans Neural Netw Learn Syst 32(5):2040–2053

Fan T, Wan Q, Ding Z (2024) Topology switching for optimal leader-following consensus of multi-agent systems. IEEE Trans Circuits Syst II Express Briefs. https://doi.org/10.1109/TCSII.2024.3372022

Guo H, Meng M, Feng G (2023a) Mean square leader-following consensus of heterogeneous multi-agent systems with Markovian switching topologies and communication delays. Int J Robust Nonlinear Control 33(1):355–371

Guo H, Meng M, Frng G (2023b) Lyapunov-based output containment control of heterogeneous multi-agent systems with markovian switching topologies and distributed delays. IEEE/CAA J Autom Sinica 10(6):1421–1433

Hu T, Liu X, He Z, Zhang X, Zhong S (2022) Hybrid event-triggered and impulsive control strategy for multiagent systems with switching topologies. IEEE Trans Cybern 52(7):6283–6294

Huang L, Xu S (2023) Impulsive stabilization of systems with control delay. IEEE Trans Autom Control 68(3):1784–1791

Liu ZW, Zhou H, Guan ZH, Feng Y, Ding L (2013) Distributed consensus tracking with reduced interaction via impulsive algorithms. In: Chinese control conference, CCC, pp 7263–7267

Liu M, Zhao L, Chen J, Yang W (2022) Formation tracking for second-order multi-agent systems without velocity measurements via impulsive algorithm subject to input delay. In: Proceedings - 2022 37th youth academic annual conference of chinese association of automation, YAC, pp 370–374

Ma T, Li K, Zhang Z, Cui B (2021) Impulsive consensus of one-sided Lipschitz nonlinear multi-agent systems with Semi-Markov switching topologies. Nonlinear Anal Hybrid Syst 40:101020

Sun J, Chen G, Chen J (2020) Stability analysis of aperiodic sampled-data systems: a switched polytopic system method. IEEE Trans Circuits Syst II Express Briefs 67(6):1054–1058

Wang J, Yang S, Wang Q, Ji L (2023) Finite-time consensus of nonlinear delayed multi-agent system via state-constraint impulsive control under switching topologies. Nonlinear Dyn 111(13):12267–12281

Wang X, Wu H, Cao J, Li X (2024) Dynamic event-triggered boundary control for exponential consensus of multi-agent systems of impulsive PDEs with switching topology. Inf Sci 655:119901

Yaghoubi Z, Talebi HA (2020) Cluster consensus of fractional-order non-linear multi-agent systems with switching topology and time-delays via impulsive control. Int J Syst Sci 51(10):1685–1698

Yang X, Li X, Lu J, Cheng Z (2020) Synchronization of time-delayed complex networks with switching topology via hybrid actuator fault and impulsive effects control. IEEE Trans Cybern 50(9):4043–4052

Yu Z, Wang X, Zhong S, Yang H (2023) Impulsive control for one-side Lipschitz nonlinear MASs under semi-Markovian switching topologies with partially unknown transition probabilities. Nonlinear Anal Hybrid Syst 48:101336

Zhou H, Wang ZH, Liu ZW, Hu W, Liu GP (2014a) Containment control for multi-agent systems via impulsive algorithms without velocity measurements. IET Control Theory Appl 8(17):2033–2044

Zhou H, QinYZ Feng K, Hu WS, Liu ZW (2014b) Impulsive consensus tracking of multiagent systems with quantization and input delays using position-only information. Math Probl Eng 2014:781048

Chapter 9
Conclusion and Prospects

9.1 Conclusion

Based on the work of previous researchers, this book focuses on the research of multi-agent systems (MASs) and focuses on the problems and shortcomings in impulsive coordination control of MASs. The main innovative research work completed is as follows:

1. Aiming at the problem that the control time cannot be accurately determined when applying impulsive control due to the limitations of environment and hardware in practice, the variable impulsive control protocol is designed. Consider each time window containing multiple impulses, and impulsive control can be applied at any time within the time window without the need to fix any impulsive instant. In addition, the proposed variable impulsive control method is extended to intermittent impulsive control, allowing MASs to not transmit information and apply control for a period of time, increasing the flexibility and practicality of impulsive control.
2. A new distributed asynchronous impulsive control strategy is designed, which only needs partial agents to transmit information and be controlled at each impulsive instant. In addition, in order to avoid the smaller and smaller impulsive interval obtained according to the inequality conditions, two improved schemes are proposed to reduce the conservatism of the asynchronous impulsive control, so as to avoid the situation that the obtained impulsive interval is too small or even negative and ensure the feasibility of the asynchronous impulsive control scheme designed in this book.
3. The problem of consensus for stochastic perturbed delayed MASs is investigated. Based on the Lyapunov stability theory, impulsive differential equations, stochastic differential equations, graph theory, and some novel delayed differential inequality techniques, some new sufficient conditions to achieve consensus

Z. Zhang et al., *Distributed Impulsive Coordination of Multi-Agent Systems*, Intelligent Control and Learning Systems 15, https://doi.org/10.1007/978-981-97-6262-0_9

are established. Since stochastic disturbances and other channel noise are ubiquitous in both natural and manmade systems, the consensus control scheme is more reasonable for the practical system.

4. The adaptive impulsive consensus control of uncertain MASs is studied for the unknown parameters and stochastic perturbation in the system model. Aiming at the problem that the continuous adaptive control law needs to measure the state and error information of the system continuously during the whole control process, a discontinuous adaptive impulsive control scheme is proposed. The adaptive law and the impulsive control protocol which only need to sample the information at the control instant are designed, and the sufficient condition for realizing the consensus of the uncertain MASs under stochastic perturbation is obtained. The discontinuous adaptive law designed only transmits information at the impulsive instant, so it is more in line with the advantage that the impulsive control can reduce the information transmission cost.
5. The consensus problem of fractional-order nonlinear MASs via distributed impulsive control method with control gain error is studied. Some new sufficient conditions are derived for the consensus of fractional-order MASs. Due to the special advantage of impulsive control method different from the continuous one, and the impulsive control error is considered, the impulsive consensus problem of fractional-order MASs is more challenging and significant.
6. The distributed impulsive control strategy against DoS attacks and deception attacks is designed for MASs. (1) Aiming at the problem that the current research results on impulsive consensus of MASs under DoS attacks are not comprehensive and conservative, the control process is divided into normal window, attack window, and regulation window based on the recovery mechanism of DoS attacks. In addition, this book is compared with other relevant results, which shows that the impulsive control method proposed in this book has wider applicability and less conservatism. (2) In view of the current research results on the impulsive consensus of MASs under deception attacks, this book considers that there are two types of deception attacks in the system, namely injection attacks and replacement attacks, and the probability of each agent being attacked is different. Different from the existing research results, this book focuses on the influence of the convergence rate of the error system and gives the selection rules of the convergence rate, the impulsive control gain, and the upper and lower bounds of the impulsive interval to obtain better control effect.
7. The time-delayed impulsive algorithm is proposed to achieve the formation of MASs with switching topologies. Unlike continuous control algorithms, the proposed impulsive control algorithm only requires applying impulsive control inputs at fixed impulsive instants and can adjust the size of the control inputs based on the information of communicable neighboring nodes to achieve the purpose of formation control. The use of impulsive control algorithm can not only reduce energy consumption, but also greatly improve the robustness of the system and

maintain confidentiality in specific situations. Based on the Hurwitz stability theory and the lemma about matrix spectral radius, the conditions to complete formation tasks of MASs are derived.

9.2 Prospects

This book focuses on impulsive coordination control of MASs and designs corresponding distributed impulsive controllers for different systems and practical situations. Of course, the control method proposed in this book also has some limitations and shortcomings, and there are still many issues that need further improvement and research:

1. The MASs considered in this book are all homogeneous systems, meaning that the models of all agents in the system are the same. However, in practice, there are often situations where the system models are different. The distributed impulsive control method designed in this book for homogeneous systems cannot be directly applied to heterogeneous systems. Therefore, it is necessary to study the impulsive coordination problem of heterogeneous MASs.
2. This book considers the situations in which the MASs are subjected to two types of attacks (DoS attacks and deception attacks), but in practice, the MASs may also suffer from both types of network attacks simultaneously. Therefore, considering the MASs are subject to both DoS attacks and deception attacks, how to design a distributed impulsive control strategy and maximize the advantages of impulsive control is also a problem worthy of in-depth research.
3. All the research content in this book only focuses on the theoretical research of impulsive control in MASs, lacking practical engineering applications. Therefore, how to apply the research results of this book to practical engineering systems (such as multi-spacecraft systems and smart grid systems), linking theory with practice, is a very meaningful and necessary work in future research.

SPRINGER NATURE

GPSR Compliance

The European Union's (EU) General Product Safety Regulation (GPSR) is a set of rules that requires consumer products to be safe and our obligations to ensure this.

If you have any concerns about our products, you can contact us on ProductSafety@springernature.com

In case Publisher is established outside the EU, the EU authorized representative is:

Springer Nature Customer Service Center GmbH
Europaplatz 3
69115 Heidelberg, Germany

The manufacturer's authorised representative in the EU is Springer Nature Customer Service Centre GmbH, Europaplatz 3, 69115 Heidelberg, Germany. If you have any concerns regarding our products, please contact ProductSafety@springernature.com

Printed and bound by CPI Group (UK) Ltd, Croydon, CR0 4YY
30/09/2026
02225308-0019